The Farmer's Guide to Astronomy

A nonfictional essay of facts and theories about astronomy
and space exploration by B.E. Chenault

B.E Chenault

authorHOUSE®

AuthorHouse™
1663 Liberty Drive
Bloomington, IN 47403
www.authorhouse.com
Phone: 1 (800) 839-8640

Published by AuthorHouse   09/13/2016

ISBN: 978-1-5246-1986-2 (sc)
ISBN: 978-1-5246-1985-5 (e)

Library of Congress Control Number: 2016911777

Print information available on the last page.

Any people depicted in stock imagery provided by Thinkstock are models, and such images are being used for illustrative purposes only.
Certain stock imagery © Thinkstock.

This book is printed on acid-free paper.

Because of the dynamic nature of the Internet, any web addresses or links contained in this book may have changed since publication and may no longer be valid. The views expressed in this work are solely those of the author and do not necessarily reflect the views of the publisher, and the publisher hereby disclaims any responsibility for them.

The purpose of this work is to engage the human imagination about the wonders of space exploration and the possibilities of traveling to its outer reaches while emphasizing the human contribution to astronomy. This book is meant to share knowledge and inspire thought in people who know little about astronomy and are too busy to research the subject for themselves. This work is collected for your convenience, entertainment and education. I don't put forth anything new, nor attempt to take credit for anyone's work. It is likely many readers will know some of this material but doubtful anyone will know all of it.

The strange dichotomy of astronomy is it's the oldest science that produces the newest discoveries. For example, the age of the universe changed in less than a year, when I began writing this book its estimated age was 13. 8 billion but by the time I finished this narrative essay NASA estimated the universe to be 14 billion years old. So not only does astronomy evolve before our eyes it's also one of the few disciplines where amateurs can make significant contributions. We'll learn how a person in space can walk at 18,000 miles an hour, discuss how a planet on fire can become a ball of burning ice and how a day can last a year. The latest research will be examined on a planet made from diamonds, you'll learn what it was like to fall through the atmosphere at more than 500 miles an hour as a person breaks the sound barrier without a jet. Find out how vulnerable Earth is to asteroids and how something called the blind spot allows meteors to strike Earth with no warning. Understand how Friday the 13th, in the near future could be the unluckiest day for millions of Americans. Read about the two only cases where people were actually struck by asteroids and how it changed their lives one for the better and the other for the worse. Absorb details of the first space disaster in which the victim speed toward Earth crying and cursing engineers who built the craft seconds before he was killed.

There's a lot material to cover in a short period so we'll move through topics fast and keep interest high.

We'll explain why there really is no up or down, only gravity. Watch science fiction become science after scientists actually catch a speeding comet. You'll be reminded that while you are standing still you are traveling 1,000 miles an hour in one direction, speeding along at 65,000 miles an hour in another direction and traveling about 250,000 miles an hour in yet another direction, again all at break neck speeds in different directions at the same time at a complete stand still. Find out why Pluto is no longer a planet.

Learn about a giant underground machine on Earth that looks like it belongs in the latest James Bond movie, it propels particles almost as fast as the speed of light in huge tunnels that pass under a number of European countries. Find out about early space travel when people attempted to parachute from the edge of space 15 miles up. Discover some of NASA's most embarrassing moments, like the time when it erased its only tapes of the first ever moon landing and how it avoided admitting the mistake for decades.

Read about the man who made air travel safer by rocketing across the ground as fast as a jet in order to prove humans could survive extreme g-forces and how NASA wanted to make space travel accessible to the masses but the plans might have been derailed by the Challenger disaster. Read about some of the people who were killed during that explosion and what they wanted to accomplish. Space travel might soon be coming the public if it can reduce ticket prices from more than a million dollars per seat to something more affordable. Learn about how metals in space automatically stick together as tight as any weld but without heat and ponder the futures of stars that are a thousand times larger than our sun. Examine some of the latest theories about light, how it can travel so fast at 186,000 miles per second and why it is the only true time traveler known to science.

Apophis's close call!

Apophis is an asteroid that could hit Earth in 14 years and kill millions of people so this might be our wake up call to try and divert it or it might be too late and we already out of time. Here's what has happened in the past when an asteroid the size of a bus or a city or somewhere in between, slammed into Earth with such speed that it went right through the rock surface some 10 miles down. The molten rock from the mantle is thrown high in the Stratosphere, circling the globe and eventually rains fire and burning rock back down to Earth. If you are close enough to see the blast the bright light and intense heat might kill you. If you are on the other side of the globe you'll probably be incinerated, not burned but incinerated by the super-heated shock wave that sweeps over the globe. Initially you wouldn't hear anything because the asteroid is traveling much faster than the speed of sound. Sound travels at 768 miles an hour; asteroids can travel much faster than 20,000 miles an hour. Science fiction or science fact! That description is now the dominant theory on how the dinosaurs were killed some 65 million years ago when an asteroid about 6 miles wide slammed into the Yucatan Peninsula and changed the planet forever.

Apophis is much smaller but more dangerous because now it might kill millions of people unlike 65 million years ago when humans and civilization had not yet evolved. There could be more than two million asteroids out there in the solar system. Since 2007 as many as 40,000 asteroids have been cataloged but there is only one that could impact Earth in the next 14 years and its name is not Flower or Bambi. It is Apophis, named for the god of darkness and destruction. Timing is everything here. Apophis was first discovered in the summer of 2004 and made a close approach to Earth just days before the Indian Ocean Tsunami struck. That natural disaster killed 230,000 people on December 26, 2004. You probably remember the day the tsumani hit but recall nothing about Apophis. That's because the tsunami made the news of the day and not Apophis which passed by Earth days earlier. While Apophis is not the size of the asteroid that killed the dinosaurs it is big enough to kill millions of people if it impacted Earth. Scientists say it is highly unlikely Apophis will strike Earth but they fail to say an asteroid's trajectory is so fragile that even sun light can change when and where they are projected to hit.

When it comes to asteroids speed makes up for size. Apophis is small enough that it could fit snugly inside a football stadium but traveling 23-thousand miles an hour it could devastate Northern California, its neighboring states and Hawaii by its impact and its ensuing tsunami. Northern California is just one of several possible impact locations.

Those who are superstitious may not like this. The next time Apophis comes close, Earth might not be so fortunate; that date is predicted to be April of 2029 on Friday the 13th. That's right Friday the 13th fourteen years from now, or more accurately when this material was written in 2016. There is a chance that this 25-million-ton space rock will affected by Earth's gravitational pull on its next flyby. The big rock could then hit us, all though scientist again say the chances are small. If Apophis passes us by, it is projected to threaten Earth with another impact seven years later on another Friday the 13th this time in 2036. Go figure, two possible dooms days and both fall on Friday the 13th, one on 2029 and the other one 2036.

If the predictions turn out to be true, 14 years from now Apophis will come an astronomical stone's throw from Earth just about 20,000 miles away. That is many times closer than the moon. At this distance Apophis could be seen with the naked eye. According to the asteroid's projected trajectory it will pass well under our satellites. In the early morning hours on that day people in Europe, Africa and western Asia will be able to see a dim object cruising westward through the constellation of Cancer. Not long after that its light will grow will dimmer and eventually vanish into the darkness of the early morning sky. But if Apophis strikes Earth it could create an explosion 100,000 times more powerful than the Hiroshima bomb. The experience of surviving the first close call could leave us breathing a little easier knowing that we dodged a bullet, a cosmic one, for the time being until it comes back around in another seven years on another Friday the 13th.

Is there Anything to Friday the 13th?

What are the chances that something bad will happen on Friday the 13th? We just found out Apophis could come calling on two Friday the 13ths just seven years apart possibly making for one bad Friday the 13th. Space junk also just fell out of the sky on Friday the 13th in November of 2015. The superstitious might not be the only people now starting to wonder about Friday the 13th. Space junk is the term used to describe all of the debris left behind from past space missions such as satellites, probes and human flights that remain in orbit until they fall back down to Earth. Friday's space junk was called WT1190F so what are the chances that it would fall out of the sky on Friday the 13th, well no one really knows.

Friday the 13th often refers to a fear of the number 13 and can be called Frigga triskaidekaphobia which is the Greek term describing superstitious behavior motivated by the number. Friday the 13th was once also called Black Friday, named well before people were trampled in Walmart. Friday the 13th or Black Friday could date back to the Middle Ages, rising from the story of Jesus and the 13 individuals who were present the night before his death. In western superstition Friday the 13th occurs in the Gregorian calendar for falling on the 13th day of the month.

According to some researchers an estimated 800 million dollars is lost every year around the world because people believe the day to be unlucky. So they don't go to work, they don't go to school, they don't shop and they don't travel. No doubt these are the same folks who avoid black cats and 13th story floors. But there's no statistical evidence that proves or even suggest that Friday the 13ᵗʰ is unlucky. Tell that to those who follow Apophis, or to people in Sri Lanka which was the best vantage point to view that falling space junk on Friday the 13ᵗʰ in 2015.

That space junk was called WT1190F and could have been some of the left over parts and fragments from the Apollo missions from 1961 to 1975 or other more recent operations. It could also be an exhausted rocket from one of the robotic missions to the moon possibly first launched from Earth just a few years ago. Like many space phenomenon estimates take the place of precise calculations. WT1190F was first identified in 2013 speeding around Earth. Its orbit was extremely elliptical which brought it close and then far out into space. These elliptical orbits can be very unstable given the nature of the routes they travel, extreme gravity when they are close and very little gravity when they are far out. At its farthest approach WT1190F was about 476,000 miles from Earth or about twice the distance of the moon. At its closest it was about 155,000 miles, which is about the mileage on many middle aged family sedans. Were told most of WT1190F burned up in the atmosphere and there was no chance of any small debris impacting the planet which was fortunate as we'll explain in the next paragraph.

So while a possible doomsday is Friday the 13ᵗʰ, fourteen years in the future, space junk is a problem right now that threatens to become worse. A previous occurrence that got the most attention was the Sky Lab incident in 1979. Debris from that larger craft reached Earth in part because Sky Lab weighed more than 75 tons. Smaller objects often burn up during reentry. Chances of someone on Earth getting hit by a piece of debris from Sky Lab were alarmingly high estimated at 1 in 152. So scientists tracking the space debris called WT1190F say that particular Friday the 13ᵗʰ in 2015 brought them good luck because all that orbital debris reentered the atmosphere without incident.

Three in One!

While America lays claim to the worst space disaster the Russians had the first one on April 24, 1967. Cosmonaut Vladimir Komarov would be the first human to pay with his life in the name of space exploration. His story is really three stories contained in one. Vladimir Komarov was admired for his perseverance but that perseverance could have killed him. He as was an engineer, test pilot and cosmonaut that was twice deemed unfit for the Russian space program. But he would not be deterred by rejection and persevered to achieve his professional goal of entering space.

He entered the program in 1960, called the Air Force Group One and he was one of its most experience and qualified participants. He was killed seven years later in what has been called a suicide flight. Vladimir Komarov's last moments speeding toward Earth were spent cursing Russian engineers as he cried in a fit of rage alone in his capsule. Even though his body was unrecognizable after impact it was put on public display. Here are some of the details of the story that are almost too bizarre to be true.

During that doomed flight Vladimir Komarov struggled with the ailing craft from the beginning, power was compromised, antennas did not work and at best there was only partial navigation. He orbited Earth several times in an effort to correct the many problems. He had to orient the craft but in order to do that he needed to see the sun which he could not. He was not left with many options and began to reenter Earth's atmosphere after his 19th orbit. The craft's main breaking parachute failed to deploy. It crashed to the ground.

Vladimir Komarov was killed upon impact. Years later there were reports that listening devices picked up radio transmissions of Vladimir Komarov's last words. He was in a rage. As he sped to Earth he was crying, apparently cursing the engineers who had built the space ship that would turn out to be his coffin. There are reports that he said the temperature was rising and he might have even used the word "kill" possibly a reference to what had been done to him.

Vladimir Komarov's remains were displayed for public viewing and were unrecognizable. The Soviets never explained their reasoning for showing the ghastly sight. The remains were displayed in an open coffin and they amounted to little more than a disfigured charcoal briquette that measure two feet in length. After viewing such a hideous sight one can only hope that Vladimir Komarov lost consciousness before he was incinerated beyond recognition. Scientists say his body turned molten upon impact. There are no indications that his craft was able to slow down and likely crashed at full speed into Earth at more than 15,000 miles an hour.

Years later one of his close friends who was also a cosmonaut said Vladimir Komarov knew that space craft was unsafe and he was probably going to die in it. Remember, Vladimir Komarov was an excellent engineer with extensive mechanical insight. But he did not back out because his close friend and fellow cosmonaut would have been selected as his replacement in the flying death trap.

So it seems not so much in the name of national pride as it was respect and a deep caring for a close friend that Vladimir Komarov did not back out of a doomed space trip but instead willingly boarded the suicide vessel with full knowledge that his trip was likely just one way. Vladimir Komarov was just 40 years old. Due to the technical facts he may not be the first human to die outer space, since he crashed into Earth. During his official funeral in Red Square there was a poignant picture of his widow kissing a picture of her dead husband.

A Day that lasts a Year!

About four billion years ago Earth and Venus were alike and in that beginning it was plausible that the two planets would mature to similarity. Today we know nothing could be further from the truth. We know Earth provides us with a warm habitable environment. Venus is so inhospitable that we can't even send a probe there without the apparatus burning up before it has time to send back pictures. The Russians were able to land a refrigerated craft on the surface of Mars which delayed the melting process long enough to send images back to Earth.

So why is Venus so different from Earth when the planets were so similar at birth? One theory is Venus was hit by a Mars sized planet just like Earth was but the result on Venus was very different. The theory as it relates to Earth is called Theia. Theia is the Mars sized planet that hit Earth at just the right speed and just the right angle so as not to destroy Earth but to change it for the better. The collision is thought to have happened 4.5 billion years ago. When Theia hit Earth part of Theia shot off into space and created the Moon. The other part was absorbed by our planet. Scientists theorize that before the impact, Earth might have rotated three times as fast as it currently does. That would mean an Earth day lasted 8 hours and the Sun rose three times in 24 hours instead of just once. The newly formed Moon also stabilized Earth's spinning motion on his axis. Just after impact Earth wobbled so wildly on its axis that life as we know it would be impossible. So the impact and subsequent Moon that formed slowed Earth's rotation and stabilized its movement which allowed life to develop. We will discuss more about the Moon and the Theia theory later.

The impact on Venus was more direct and at a higher speed. The collision nearly stopped Venus from spinning on it axis. As a result, it takes Venus 243 Earth days to make one complete revolution on its axis. But it takes 224 days for Venus to orbit the Sun. So one day on Venus is longer than its year.

Another consequence of the impact was Venus lost its magnetic field. Magnetic fields are generated by planets that have iron cores and spin fast enough on their axes to make the fields beneficial. With its magnetic field gone Venus lost its hospitable atmosphere. Radiation entered its environment and broke down basic elements like h20 and water evaporated. The natural chain was broken in than heat from the Sun could reach Venus but that heat could not get back out. The planet's greenhouse gases soon took over and the temperature of Venus soared to 800 degrees. Even though Mercury is the closest planet to the Sun, Venus is the hottest planet in the entire solar system with a surface temperature of about 800 degrees.

Venus has clouds made up of sulfuric acid, similar to the compounds in car batteries. While these sulfuric acid clouds produce rain, Venus is so hot that the rain evaporates before it can reach the surface. The surface on Venus is completely uniform unlike Earth's where there is water, land, ice and rock. On Venus there is only lava. That's because many large volcanic eruptions released massive amounts of lava that covered the entire surface of Venus in one uniform mat. Turns out Venus is not the goddess of love, beauty, sex and fertility as referred to in mythology, instead it is much closer to a hell, a hell that could have befallen Earth if not for mere chance.

They actually Caught a Comet!

The never say die attitude exemplified the effort from a group of scientists who succeeded against the odds and were able to land a space ship on a speeding comet. The space ship was called Rosetta and it carried the probe called Philae which would actually land on the comet. The mission was daunting enough without any problems but Rosetta could not catch the original comet that was chosen so it had to travel farther, some five billion miles in all to complete the mission using a procedure called a gravity assist to increase its speed. Gravity assists sounds like something out of a science fiction cartoon but it's scientific reality and works like this. When a space craft needs to increase its speed it flies close enough to a planet where the planet's gravity catches the space craft and flings it out into space like a sling shot which gives the craft a tremendous acceleration.

Rosetta performed four gravity assists, the first one was around Earth which almost tripled Rosetta's speed up to 76,000 miles an hour. It performed two more around Earth and the last one was around Mars before it was realigned in the right position some 62 miles above the comet. Don't know why the crew chose to fly 62 miles above the comet but on Earth, 62 miles above our planet is the area where space actually begins.

One can only marvel at the math needed for the precise calculations required to place that small probe, about the size of a washing machine in the right places at the right times, millions of miles away in the pitch black darkness of space while traveling more than 50 thousand miles an hour. But more problems lay ahead. Scientist said that Rosetta would not really land on the comet but instead kind of dock with it because of the comet's low gravity. The gravity on the comet was so low that there was a good chance Rosetta would just bounce off. To remedy this problem, Rosetta was fitted with hooks and harpoons to better secure it to the comet's surface. But those devices did not work as planed possibly due to the fact that much of Rosetta's power had to be shut off for years in an effort to save energy. That energy saving procedure meant Rosetta was put into a deep hibernation. The hibernation was required in order for Rosetta to solve a series of previous problems. After the prolonged hibernation there was no guarantee Rosetta would wake up.

Rosetta's purpose was to learn more about comets and among other things to determine if they carried certain amino acids which are believed to be the building blocks of life. But perhaps equally impressive were the difficulties of the mission, the ambitious nature of the project and the tenacious characters of the scientists involved who refused to be defeated. The story of the man behind Rosetta is almost as amazing. If you were to see this young man on the street you might guess he's a punk rocker. He has spiked hair. His fore arms are full of tattoos. He often wears heavy metal t-shirts. But he comes from a family of modest means. He labored side by side with his father in a back breaking, labor intensive occupation while he was worked his way through college. Dr. Matt Taylor is a devoted son, father, husband and complete family man and yes, he's also one of the smartest people on Earth and the lead scientist behind the Rosetta mission.

Here's a more detailed account of some of those problems. Rosetta's delayed launch meant that it had to find a new comet to explore. So the new comet had to be big enough to land on and close enough to reach. The new comet met the first criterion because it was about two and a half miles wide but it was at the boundaries of how far Rosetta could travel, at about 500 million miles from the sun. So Rosetta was put into that deep three-year hibernation to save enough energy required to make the long journey. If Rosetta didn't wake up all of the scientists who worked for years on the project would probably just go home because they would be out of jobs.

Another one of the many problems was Rosetta did not know where it was when it came out of that three-year hibernation, so it could not point itself in Earth's direction to send the message to mission control that it was alert and ready to work. So Rosetta took pictures of the surrounding stars and compared them to the ones in its catalog and after a process of elimination became oriented with its position. This allowed Rosetta to point itself toward Earth and send the much anticipated message back to Earth. So on the fateful day when Rosetta was supposed to call home, two huge satellites one on each side of the globe were pointed in the right direction, waited and listened from a signal which finally came about eight hours late. When the message was received and confirmed seasoned scientists celebrated the news with more enthusiasm than uncontrollable children bubbling with the amazement of their first Christmas morning.

So after 10 years, five billion miles from home, a three-year hibernation, traveling at 41 thousand miles an hour, this 1.7 billion dollar investment was ready to "catch a comet." Problem was there were more problems. The comet started to erupt into a gas storm and the closer it got to the sun the worse the comet storm became. The sun's rays had unlocked gasses trapped inside the comet which caused the comet's gas storm. That storm blocked its view and without a view scientists did not know where Rosetta could land on the comet. The comet looked like a doubled headed dumbbell, bigger than originally thought, full of jagged rocks, hidden crevices and big boulders. Trying to land on a surface like that was made more difficult by the comet's light gravity, thought to be 100,000 times less than Earth's. So light, that if a person were to walk on the comet that person would bounce along as they strolled and could literally jump on the surface and float away into space. The gravity was so weak that the comet might have been lighter than water and may actually float in liquid. Long story short Rosetta's lander touched down on the comet and bounced for a while but did not bounce off and eventually rested on the comet completing what arguably was humanity's finest hour in space in the year 2014.

Speeding while Standing Still!

How can we be traveling 1000 miles an hour while standing still? Better yet, how can we travel at 67 thousand miles an hour while still standing still? Topping that, how can we travel at 483,000 miles while still standing still? And ultimately, how can we be moving about a million and a half miles an hour while we stand motionless on the same spot on Earth? And finally how can we be traveling at all of these different speeds at the same time, in different directions, while standing still? It sounds impossible but ask any reputable scientist and they will tell you it is fact, common knowledge in the field of astrophysics and astronomy.

Consider the fact that our planet is spinning on its axis at about one thousand miles an hour, taking the planet about one day to complete its revolution. Now consider Earth is traveling about 67 thousand miles an hour as it orbits the Sun, taking Earth about a year to complete that orbit. Also consider that the Sun is traveling about 483,000 miles an hour in its orbit around the Milky Way. The last time the sun completed that orbit, the dinosaurs were alive some 230 million years ago. And finally the Milky Way which is our home galaxy is traveling toward the nearest galaxy, Andromeda at a speed of about one and a half million miles an hour. Put another way, we are on Earth and Earth goes around the Sun and the Sun goes around the Milky Way and the Milky Way travels toward Andromeda, they are several rides all contained in one larger ride. So there you have it, the commonly agreed upon scientific explanation as to how we can travel several different directions at differing speeds all at the same time while standing perfectly still in one spot on Earth.

On a related note, if you study the sky closely you will notice that it changes a little bit each night. Again that's because the entire universe is in motion. Each night the stars and other bright objects move a little farther to the west than they were the night before. Scientists say that's because the Sun and Earth rotate from west to east in a counter clockwise motion. They also say this makes a star rise about four minutes earlier each morning which adds up to about two hours earlier each month. So many stars that were seen last month at night now become obscured by day. After a year has passed, everything is reset, all of the stars are back where they were 12 months ago and the night sky looks the same.

The same laws that apply to the movement of stars in the sky also apply to our seasons. The sun travels highest in the sky during the summer and is lowest in the winter. The sun's highest point is called the summer solstice which produces the longest day of the year and its lowest point is called the winter solstice, producing the longest night of the year. Remember all this is possible because the entire universe is in motion.

This is easier to understand if you've ever seen a computer model of this phenomena. One picture really is worth a thousand words. Picture a computer model of the solar system and galaxy as a series of rings where one ring is contained inside the other and the outer ring is the largest and the inner ring is the smallest. The outer ring is our galaxy, the Milky Way and it is spinning in one direction. A middle inner ring is our solar system with the sun at the center and it is traveling in another direction. The smaller inner ring is Earth traveling as it orbits the Sun. And the smallest ring would be the lone Earth as it spins on its axis. They all might not be traveling in completely opposite directions but they are traveling in different directions.

If there is intelligent life in the universe it could likely experience the same "backward forward at the same time" phenomena. According the latest information there are an estimated 100 billion galaxies in the universe and that number is expected to go up as space telescope technology improves. So if half of those galaxies are like the Milky Way there could be an estimated 50 billion of these "backward forward at the same time" phenomena occurring in the universe. Breaking it down even further, if there are an estimated 3 trillion solar systems in the universe and half of those solar systems are home to intelligent life then there could be 1.5 trillion other beings who experience this concept.

So at the smallest level we have the planet, which makes up the solar system and the solar system makes up the galaxy and the galaxy makes up the universe and they all are moving at the same time, in different directions and at varying speeds, which are connected to the movements of the stars and the passing of the seasons and all of it is happening overhead at super-sonic speeds while we stand still and watch in amazement.

The T is Silent!

When asteroids crash into the ocean they most likely cause giant tidal waves called tsunamis. The size of the asteroid determines the size the tsunami, some waves are dozens of feet high others are hundreds of feet tall. These giant waves can also be caused by earthquakes but we will just focus on those caused by asteroids. One of the problems identifying an asteroid that caused a tsunami is finding the impact crater it left behind which is usually submerged under thousands of feet of ocean water. So scientists look for other signs that are more visible, such as something called a chevron.

A chevron could be described as a small mountain or a very large hill with a V shape. They're often left behind from tsunamis that were caused by asteroids. Chevrons are usually miles from water after a huge wave deposited them to their present locations, the force of the rushing water forms the chevrons into their V shapes.

There are four of these chevrons on Madagascar's south side. They are about three miles from the ocean and each are some 600 feet high. They contain organic material from the ocean including small fossils which fused together with metals commonly formed after asteroid impacts. All of the chevrons point toward the Indian Ocean which is a telltale sign of their origins. Turns out after scientists followed these clues they found an impact crater 18 miles in diameter some 12,500 feet below the ocean surface. They believed the asteroid crashed almost 5000 years ago, it was a blast so strong that had it happened today it might have wiped out 25 percent of the world's population. It produced a tsunami estimated to be 600 feet high which would be more than a dozen times as large as the tsunami that crippled parts of Indonesia in 2004. So the mystery seems to have been solved of how those chevrons were formed about three miles from the ocean with heights of 600 feet.

The new discovery in Madagascar threatens some old beliefs about how often asteroids impact Earth. One of the more commonly held views is that there have not been any major asteroid impacts in the last 10,000 years. But there's evidence of one in Madagascar over the last five thousand years. In fact, this evidence could overturn the commonly held belief that Earth is impacted by asteroids once every 500,000 to one million years. Now that frequency rate could be reduced to once every ten thousand years.

So scientists could have been looking for the "smoking gun" in the wrong place and have now refocused their efforts. Searches for asteroids that cause tsunamis now center along coast lines and scientists look at chevrons in a whole new light. With the use of the latest technology, chevrons could be found across the globe which is both good and bad. Good because science will certainly benefit from the knowledge, bad because the chances of getting hit by an asteroid just went up since they might strike far more often than what was previously thought. Chevrons have now been found in Australia, Africa, Europe and the United States including New York City. All of the chevrons point in the same direction, toward the ocean, as if they were trying to tell us something or leave a clue as to their origins.

One common school of thought is about 175 large asteroids have hit Earth in the distant past and most of those impacted land and not the ocean. As a result, most of the searches were organized on land and not underwater. A new method could bring about a change in the way scientists look for asteroid impacts but the new method is not without problems.

One difficulty is how does one look for evidence of an asteroid impact thousands of feet underwater on the ocean floor? The crater impact would most likely be filled in with sediment. Part of the solution is to use satellites that scan the ocean surface and then determine its depth. Then scientists measure the distance from the ocean floor to the ocean surface which helps them narrow the focus of their search. Then they eliminate underwater mountain ranges, trenches and sink holes and whatever is left could be an impact crater from an asteroid. Of course not all depressions in the ocean floor are impact craters from asteroids but once the search is sufficiently narrowed, divers can go down and take soil samples of the underwater crater to see if it contains cosmic material like iridium. Iridium is common in space but rare on Earth. Newly discovered underwater impact craters from asteroids have already been found using this method, which again could be good or bad news depending on your point of view. But what it does mean is that Earth might get struck by an asteroid once every ten thousand years instead of once every million years.

The power and destruction of some of the ancient tsunamis is much more impressive than what is demonstrated by some of the more recent ones. Can you imagine an ancient tsunami so big that it covered the globe, everything except the highest of elevations was submerged? Scientists say they have found evidence of such an impact which could also be the earliest known asteroid strike, believed to have struck 3.5 billion years ago. The giant tsunami could have rushed around the globe several times and again covered everything except the highest mountains.

3.5 billion years ago Earth's continents had not yet broken off into the separate pieces that we recognize today. Also at that time ocean water covered more land than it does today. Scientists say they found traces of this strike in some of the oldest rocks known on Earth in South Africa and northwest Australia. The two continents were still joined at the time of the asteroid impact. Whatever life lived on land or in water was probably wiped out with the possible exception of the smallest of animals that might have lived deep within rocks or tucked away some place safe in the ocean.

Scientists also say that tsunami swept around the world at several thousand miles an hour and when its wave hit the opposite of ground zero on the other side of the globe it reversed its direction and rushed around Earth in the opposite way. The wave repeated this motion several times, which took it about 30 hours to circumnavigate the 24-thousand-mile distance around the globe. One can hardly imagine how much energy it would have taken to move that much water, over that great distance at that tremendous speed. Dozens or maybe even hundreds of feet of the ocean's upper water levels might have been evaporated in an instant after the impact. The asteroid could have been as large as 20 miles in diameter. An asteroid one mile in diameter might be large enough to cause global extinction, let alone one 20 times that size. Often times when a large asteroid impacts Earth, debris is ejected into outer space and ends up on the Moon or Mars and apparently that was the case 3.5 billion years ago when the largest asteroid known yet struck the planet.

It has been argued that one of the most destructive forces on Earth is water, which might only be surpassed by an asteroid impact, the combination of the two thought to have occurred 3.5 billion years ago with the culmination of the giant tsunami might have been the largest disastrous occurrence Earth has ever endured.

Will the Sky Fall?

Most people probably don't give something like orbital debris much thought unless of course they happen to work for a satellite company or with a government space program. And while recent Hollywood movies have attracted attention to the problem big screen producers have apparently blown the situation out of proportion. Orbital debris is all the junk left behind from years of space flights by numerous countries and corporations. The official definition is something like "all manmade objects that orbit Earth that no longer serve a useful purpose." That could be anything from broken off nuts and bolts, to entire satellites broken apart after collisions or explosions, to metal parts intentionally released during the separations of those crafts, to even tiny pieces of paint broken off from the stress of extreme heat.

When you include all of the space debris in orbit, there could be up to a million pieces flying overhead and about half of them, an estimated 500,000 pieces could be larger than two inches. Beware! Because what they lack in size they make up for in speed. Most of the debris came from satellites that either exploded, collide with each other, or were decommissioned. It is estimated that the vast majority of this debris is hurtling around in orbit at up to 17,000 miles an hour. You wouldn't stand much of a change surviving an impact at that speed no matter how small the debris.

How long this debris remains in space depends on the height of its orbit, the higher they are the longer they stay up. Debris in orbits of 40 miles up usually will stay there for no more than a few years before they come back down. Debris at twice that height could stay up for a 100 years or more. Much of the debris that falls out of orbit never makes it all the way back down because it is incinerated in the extreme heat of reentry. The debris that does make it all the way down will probably not hurt anyone because they'll likely fall into water, understandable considering water covers more than 70 percent Earth's surface. According to one report, even though space debris has fallen to Earth every day for the last 50 years there are no credible reports of injury to people. Several government agencies have taken note of the growing situation of orbital debris and have taken the first minor steps to mitigate the problem but currently there are no international treaties to remove orbital debris.

In 1996 a French satellite was hit and damaged by orbital debris left behind from a rocket that had exploded 10 years earlier. The collision between the two space craft resulted in an additional 465 pieces of space debris. That collision also illustrates another problem: when debris collide with each other they increase their numbers instead of reducing them. Scientists report that France is responsible for about 3 percent of the debris in orbit. France is the fourth largest contributor to the orbital debris problem.

China is the 3rd largest contributor to space junk. In 2007 China intentionally destroyed one of its weather satellites. When that satellite was destroyed it added another 3,000 pieces of debris to the orbital junk pile. The destruction of that Chinese satellite could have been the single largest contributor to debris in Earth's orbit. We are told debris from China makes up about 22 percent of the junk pile in Earth's orbit.

Russia was the first orbital offender and remains the largest to this day. Remember the first craft in space was Sputnik, Russian for "satellite" and it is believed that about 37 percent of all orbital debris originated from Russian crafts.

The United States is the second largest orbital debris offender. In the winter of 2009 it was involved in one of the worst orbital collisions on record. An American satellite was destroyed when it collided with a Russian satellite. The crash occurred about 500 miles up and as they always do, it left more debris in its wake to orbit Earth and potentially crash into even more objects. The two satellites crashed into each other at about 26 thousand miles an hour. That high speed collision left behind at least 1,000 pieces of debris that were larger than 4 inches, again all of that junk will eventually increase in numbers after more collisions.

It is estimated that about half of all litter on US road ways was put there intentionally. The reasons for littering are many: it was easier to litter than to dispose of the waste properly, people don't feel responsible for the areas they litter because they see other people doing the same and the more they litter the more it becomes a habit. Is there a connection between human behavior and national behavior?

Once the Russians left space junk behind other countries saw what was done and followed suit. Was it not easier for the Russians to engineer its Sputnik craft to crash back into the atmosphere than it was to develop a more environmentally sound method of disposing of the capsule? That question can be directed to any country or corporation that sends rockets into space.

Do any nations feel responsible for littering space? The answer might be found here on Earth. It certainly seems our littering behavior will continue as we explore space. If there is intelligent alien life in the galaxy watching us what kind of message are we sending? Put another way, who in their right mind would want a neighbor who is a habitual litter? This kind of behavior has serious consequences for communication, entertainment, navigation and security systems on Earth which are outlined in something called the Kessler Syndrome in later chapters.

More Impressive than the Sun!

Imagine a star much smaller but yet a lot heavier than the sun spinning at 600 rotations per second. The sun spins once a month, or more accurately, one rotation every 27 days at its equator and once every 37 days at its polls. The reason for this is the Sun doesn't have a hard, solid surface like Earth, so the Sun's equator spins faster than its polls. The same thing would happen if you were to spin a water balloon, the middle part would rotate faster than the ends. So one star, the Sun spins about once a month and the other star spins about 600 times a second. This comparison should give you a better idea just how fast a neutron star can spin. Problem is, the human brain may be incapable of appreciating that kind of speed.

Neutron stars normally rotate about once every 60 seconds which is a more comprehend-able figure. Imagine our sun spinning on its axis 60 times a second instead of once a month. Large objects spinning at this speed can have tremendous mass and gravity. Scientist say a cup full of material from a neutron star would weigh as much as the moon. The moon might weight much more than 81 billion tons. A cup weighs a few ounces. How is this possible? Neutron stars are the result of a massive collapse that is an exercise in material condensed beyond human comprehension, which was demonstrated by stating the moon's weight. Most people just cannot appreciate a number that large, so imagine if one million Earths collapsed and were condensed into an area the size of Chicago. You now get a better idea of the term "material compressed beyond human comprehension" and why this hypothetical Chicago would be so heavy.

Obviously there is incredible pressure at the core of a neutron star, so much so that scientists say a space craft whose pilot is foolish enough to venture too close would be crushed and then made part of the star in an instant. So what is this core made of? Scientists don't know. But they do know neutron stars are one of nature's most amazing achievements. They are the densest stars that we know of in the universe. Everything about a neutron star seems to dwarf any characteristic or trait of the Sun or Earth. The magnetic field of a neutron star is much stronger than Earth's' by at least a quadrillion. A quadrillion contains 15 zeros compared to a million which has six. Scientists also tell us if Earth came within a 100,000 miles of a neutron star its magnetic field would destroy data on every credit card on Earth. These stars are so impressive that they almost defy description. They give scientists an opportunity to study forces in space that in no way could be recreated on Earth. They provide a lesson in gravity and velocities that are so extreme our brightest minds can barely understand them but what little they do grasp has sent scientists in new directions that they never would have gone on their own.

Neutron stars are almost completely made up of neutrons that are packed so tightly together they helped spawn the theory, two objects cannot occupy the same space at the same time. As we will learn later, neutrons can be separated from their atoms and by themselves, neutrons can be packed more tightly together than could ever be possible with atoms, which explains their tremendous density and gravity. Not only do we not know what the cores of these are neutron stars are made of, we're also not sure about their atmospheres. Scientist theorize their atmospheres are micrometers thin. One micrometer is equal to one millionth of a meter and a meter is similar to a yard so there's not much there! Put another way Earth's atmosphere is about 300 miles thick, a neutron star's atmosphere is much thinner than a sheet of paper. Such a thin atmosphere should not come as a surprise considering a neutron star can spin so fast it might throw off the thinnest of atmospheres altogether. Another theory is that its crust is solid, extremely hard and smooth, made this way by its intense and extreme gravitational field and the speed at which it rotates. Facts and theories about neutron stars are too long to list here but suffice it to say they helped spawn the saying science is stranger than science fiction.

Sputnik the Traveler!

The first man made satellite was launched by the Soviet Union in the fall of 1957. It was called Sputnik which means the traveler, or co-traveler or fellow traveler of this planet or with this planet or something like that. The intended definition is still debated till this day. But as best as I can ascertain Sputnik means; as the planet travels through space, Sputnik will travel with the planet as a co-traveler or companion of the planet or again something like that.

Anyway, Sputnik was about the size of a basketball but weighed as much as a human at 180 pounds. At the time it was 20 times larger than the satellite the Americans were trying to launch. Sputnik flew some 500 miles above Earth at a speed of 18,000 miles an hour. It no doubt was of great concern to the American military as it flew over the United States seven times a day during the Cold War as it orbited Earth once every 99 minutes. It carried a simple radio transmitter that sent back a series of beeps to Earth for 23 days until its batteries died. In remained in orbit for little more than a year and burned up during re-entry.

A few months before the original Sputnik burned up during re-entry Sputnik II was launched. It was much bigger than the first and weighed about a thousand pounds. Sputnik II might be best remembered for carrying the first living thing into space. It had a dog on board. The animal's name was Laika and died in space. Reason for the dog's death: there was never a plan to get Laika back to Earth. As you can probably imagine the Soviet dog in space story sparked all sorts of news reports with many of them coming from a jealous American media. Some reporters called the story Mutnik instead of

Sputnik. So much for "man's best friend" abandoned in space and resigned to certain death. The launch was supposed to test the bounds and safety for human space travel which were soon to come.

Laika was a stray dog found on the streets of Moscow less than a month before the launch. As if made any difference to Laika, she was promoted to the rank of a cosmonaut and was chosen because she was small and had a pleasant demeanor. The Russians attracted much criticism for using the dog as a guinea pig with some folks claiming the dog will die and can't be saved, which was true. Suddenly Moscow went from the celebratory mode to the damage control mode and issued a statement that Russians love dogs and that it was done not out of cruelty but in order to benefit humanity. Years later Russia was still responding to the story especially after Russian claims that the dog died a humane death. It was later reported that Laika died anything but and instead passed away just hours after takeoff into the 23-day flight from panic and overheating.

All this was at the height of the Cold War and the Soviet success in orbit put the Americans on the back foot clearly showing they could not keep pace at this stage of the space race. People thought if the Soviets could put two Sputniks and one dog into space, then could they send a missile across the ocean and hit American targets. People also thought how long would it be before Americans would catch up. All this came on the heels of something called Project Vanguard which was America's dismal attempts to launch satellites into space but ended in embarrassing failures, twice.

But as is often the case, something good can come from something bad. Many people said, what was born out of the Cold War became the Space Race. At short while later America "stepped up its game" and successfully launched its own satellite into orbit. That would be followed by the creation of something called NASA. A broader impact could be seen in the new encouragement of thousands of young Americans into fields like science and engineering in general and space exploration in particular.

Chicxulub!

Scientists say all life in the universe including on Earth was created either by an explosion or a collision. If this is true we can thank the latter for the emergence of humanity. The theory that an asteroid killed the dinosaurs is widely accepted today but was laughed at some 40 years ago. Had there been no asteroid collision, it's likely that dinosaurs would still be ruling Earth considering they had done so for 135 million years. 65 million years ago an asteroid changed all that, in fact it changed everything on this planet and for us it was for the better. At that time our mammal ancestors were so insignificant the were probably not even a food source for the biggest of dinosaurs as those small mammals were just worried about not getting crushed to death before they could evolve into humans millions of years later.

The dinosaur killing asteroid is probably one of the most interesting stories about Earth as it relates to the universe. According to the theory, some 65 million years ago the asteroid came crashing into Earth's atmosphere. Some scientists call it the Chicxulub impact named after the town in the Yucatan Peninsula where the crater was discovered. The asteroid was traveling around 25 to 30 thousand miles an hour, thousands of times faster than the speed of sound, so there was no sonic boom, you couldn't even hear it. The asteroid is some 6 miles wide and when it impacted Earth the light was so bright it might have killed you. But again you don't hear anything, just a super bright, blinding light and an eerie silence. According to theorists, when the asteroid impacted Earth the sound would have been deafening, the six-mile-wide space rock blasted a hole in the surface of the Earth 110 miles wide and 12 miles deep. In less than a

28

second the asteroid bore underground while a cloud of super-heated toxic dust, steam and ash swept across the globe. Some scientists say the shock wave would have been much worse than getting blasted by your oven when it's on self-clean. They compare it to walking outside and getting hit by an 800-degree wind that traveled at 1,200 miles an hour. That blast was much stronger than anything we can imagine. We are told the energy released from the impact was 100 million times stronger than the Hiroshima bomb. That bomb killed 80,000 people instantly and leveled major city in the blink of an eye.

The Chicxulub asteroid created some of the largest tsunamis on Earth including the giant wave that confirmed this entire theory of how the dinosaurs were killed by a huge rock from space. That tsunami was estimated to be about a thousand feet tall and swept over Florida like the state was a giant sand bar. The tsunami then deposited iridium and other materials from the asteroid off the South Carolina coast. Remember iridium is common in space but rare on Earth. It was the combination of the iridium and the discovery of the submerged impact crater in the Yucatan Peninsula that confirmed the theory of how the dinosaurs were killed by an asteroid.

The excavated Earth and parts of the asteroid were shot out into space and when they fell back down they would have been incandescence, hot enough to heat iron where it glows white. The super-heated material fell from the sky caused global wild fires. The extremely hot shock waves caused earthquakes and volcanoes. One thing built upon the next and the planet was turned into an oven in a very short time. Earth was covered in a cloud of super-heated dust and ash for years, maybe a decade. It was a massive greenhouse effect. Cooling would eventually follow as sunlight was prevented from reaching Earth's surface because of all the dust and ash still in the atmosphere. Life on Earth was changed forever.

So when and where was all this discovered? Like so many important discoveries this was one was found by chance by a team of Mexican geophysicists looking for oil with the Pemex company in the Yucatan Peninsula. They came across what appeared to be half a circle buried beneath the water and upon closer inspection they came across the second half of the circle. They then deduced that the two half circles could be put together to form a larger compete circle or a crater from an asteroid impact.

The idea was proposed in 1980 by a father and son team of geologists, Luis and his son Walter Alvarez. Years earlier the Pemex corporation had collected data suggesting such an impact had occurred but the oil company kept the information to itself. The Alvarez's became suspicious in the late 1970's when the area contained large amounts of iridium. Iridium is almost always left behind from asteroid impacts and it helped to confirm the theory a few years after it was proposed.

Remember those stories of old about balls of fire raining from the sky and how the meek shall inherit the Earth. Apparently those events actually happened millions of years before they were imagined and passed down through the generations in the form of religious warnings. Balls of fire rained down from the heavens sounds like a passage out of the Bible instead of the work of a paleontologist piecing together what that aftermath might have been like some 65 million years ago. Soon thereafter Earth's surface became so hot that rock melted, it is thought that any living thing bigger than your fist would have perished, the first animals: incinerated, the later animals died a more painful and prolonged death, which meant the small mole like animals that could burrow underground might have survived, hence "the meek would inherit the Earth." Life in lakes and oceans probably would have suffered the latter fate, first being insulated in the water, but soon after the foundational food chains such as plankton would break down and die, triggering a much larger global food chain collapse.

Asteroid impacts like the one that killed the dinosaurs have happened before and we are told it will happen again. Had the comet Shoemaker

Levy, which crashed into Jupiter in 1994 crashed into Earth we would be gonners more than 20 years ago. Scientists estimate the occurrence of such a global disaster caused by an asteroid happens once every several million years. But you'll remember there's new evidence from chevrons that suggest Earth might get hit every ten thousand years or so by a major asteroid. It was 65 million years ago since an asteroid six miles wide visited the planet and changed its evolutionary course for ever. If an asteroid half that size came down now it would probably cause twice as much damage considering the state of current affairs, civilization, dozens of mega cities and the staggering infrastructure needed to support 7.1 billion people living on the planet, all of which did not exist when the Chixulube Asteroid struck. Some scientists say if an asteroid just one mile wide hit us we would likely get wiped out.

So as beautiful as the moon is, it is also a grim reminder of what the past has held and the future still holds. The moon's surface is full of asteroid craters, so full that some of the craters overlap one another. Quite a number of scientists say the moon's craters should serve as a wakeup call to what's coming. So the next time we gaze up at a full bright moon in the night sky, its many craters send us a message that belies its peaceful, calm appearance. But this raises the disturbing questions: what good is a wakeup call if you can't do anything about it?

Too Much Cost Cutting!

Before the invention of DVDs there was something called VCRs which were devices used to record things like movies, TV shows, weddings and other special events. People who have used them usually have stories about how one of those special events was accidentally erased or recorded over with another event that was much less important.

NASA has its own story. That's right the all impressive National Aeronautics and Space Administration, known for its deliberate and carefully planned accomplishments accidentally erased the recording of its finest moment, when it landed on the moon. That's right and it bears repeating, NASA erased the recording of its first moon landing, it's finest hour, it's ultimate achievement and it had no back up copy. Till this day, the moon landing is still probably the greatest technological achievement for humanity and NASA erased it.

If the common lay person were guilty of such an egregious mistake he or she might not admit it and NASA certainly didn't either, at least for years. In fact, it was more than 30 years after the 1969 moon walk before NASA even admitted it had a problem. Seems of group of retired NASA employees tried to locate the tapes in the early part of this century. But it wasn't until the summer of 2006 when a newspaper broke the headline that brought the full story in public view. That headline read "One Giant Blunder for Mankind" which was a spoof from those famous words of Neil Armstrong, the first man to step on the moon after which he said "That's one small step for man, one giant leap for mankind."

But after enough time passed NASA admitted that it had misplaced those tapes and that they were probably somewhere in its giant video archive. The truth was coming out, a little at a time and some 37 years later all of it would be laid bare for the world to ridicule. The next step for NASA was to go beyond the "the tapes were misplaced" line and give the whole embarrassing truth.

So NASA seemed to blame the mistake on budget restraints. Turns out those irreplaceable tapes were among 700 boxes of tapes that were "misplaced" but officials were hopeful that they could be found. In 2009, three years after NASA admitted it didn't know where the Apollo 11 tapes were and 40 years after the moon landing that it was supposed to record, the National Aeronautics and Space Administration finally came out with the whole truth. The tapes were shipped to Australia where they were erased and then reused a few years later. If NASA wanted to save money, why ship the tapes half way around the world when they could have been erased much cheaper and just as easily right here in America?

Part of its official statement blamed cost cutting measures. NASA had not purchased new tapes because of budget cut backs and instead would try to reuse as many of the old tapes as possible for future recordings. So the original files from the first moon landing were reused and recorded over so NASA could save money. That's the embarrassing truth and as ridiculous as that sounds, it absolutely believable.

After an exhaustive investigation, it was uncovered that NASA had erased 200,000 tapes in order to save money. Considering what they had just been through, those investigators may not have wanted to know what was on those 200,000 tapes. One of the major networks would come to bail out NASA. That network covered the moon landing in exhaustive detail and lent its files to NASA so NASA could put its "cost cutting" house in order by finally recording the tapes some 40 years after it was supposed to.

Light Speed and Sight Speed!

Light speed seems impressive until it is compared to the universe and then attached to the short life spans of humans. The speed of light travels at 186,000 miles per second, 670 billion miles an hour, six trillion miles in a year, fast enough to go around the Earth about seven times in a second. It takes the common commercial jet liner about three days to fly around the globe. Problem is, light speed is only impressive if you live on Earth, otherwise the speed of light is inadequate for human space travel.

Andromeda is the closest galaxy to ours which is the Milky Way, the two are separated by about 2.5 million light years. It is a common belief that the best chance to discover intelligent alien life would be on a planet in Andromeda. Problem is Andromeda is so far away that there is no chance of getting there. So the concept of traveling in space at the speed of light remains just that – a concept. Many scientists might think there must be something faster than light, although publicly they state that nothing in the universe can travel faster than light. Again light speed is much too slow for humans to travel in space.

So why not consider something else, instead of "light speed" how about "sight speed." That's right, travel at the speed of sight, as abstract as the sounds, it would be an adequate means of space travel for humans. Of course this kind of travel is currently impossible but who knows what the future holds. How many people thought traveling to the moon was impossible before we actually went there. We should not underestimate the speed at which technology moves. Should scientists suggest "upping the game" and move in new directions, directions that were never considered before! It is a huge challenge but how else can we traverse the great expanse of the universe which is measured in light years when we are limited with life spans that last a fraction of that time and space craft that are much too slow. This kind of travel might lie beyond our comprehension but that does not mean it will always be out of reach.

Consider the ability to travel to a location as fast as you can see it. The moon is about 238,000 thousand miles away but we can see it in an instant, so the speed of sight might be 238,000 thousand miles per second. The sun is about 93 million miles away so the speed of sight might be 93 million miles per second. The farthest star that can be seen from Earth with the naked eye is about 2.5 million light years away so is the speed of sight 2.5 million light years per second? Light speed is universally accepted but sight speed has probably never even been considered. One of the factors that makes light so fast is it doesn't weigh anything nor does it transport any material when it travels, two facts that might help engineers improve the speed of human travel in space. Scientists have already begun to study Quantum Mechanics in the search for increased speed. These extremely small particles at the subatomic level move at amazing speeds and science has shown before that we can unlock some of the biggest mysteries of the universe by studying some of its smallest parts. We'll focus on Quantum Mechanics later in the book.

Back to reality! One of the fastest space ships ever developed by humans is likely to be the New Horizons space craft, it was the fastest craft ever to leave Earth's orbit 100 times faster than a jet. It was launched in 2006 on a journey to Pluto and then on to the Kuiper Belt. New Horizons travels around 36-thousand miles an hour. In 2004 the Rosetta space probe was launched. It could have reached speeds in excess of 70,000 miles an hour thanks in part to something called a gravity assist. A gravity assist is really nothing more than when a space craft sling shots around a planet. Here's how it worked.

Rosetta needed to increase its speed so it could catch up to a comet so engineers positioned Rosetta where Earth would pass right by. Earth travels around 67,000 miles an hour as it orbits the Sun. When Earth passed Rosetta, Rosetta was traveling around 24,000 thousand miles an hour. Earth's gravity caught Rosetta and gave it the gravity assist, a boost or sling shot and dramatically increased Rosetta's speed to about 76,000 miles an hour. Sounds like science fiction, or better yet like an abstract concept such as sight speed! No, this was reality. But before Rosetta, gravity assists and chasing comets could become a reality people first had to dream about them. I am sure those first initial dreams were called abstract concepts just as sight speed is called an abstract concept. But there can come a time when the abstract becomes the concrete.

C'mon Mr. Young!

Who would have thought that the good willed gesture of sharing a sandwich with co-workers would put all involved in a potentially deadly situation, apparently not Gemini 3 Astronaut John Young. The Gemini project had a number of firsts: the first American spacewalk, the first time two space craft were able to dock and the first potentially dangerous situation caused by the eating of an unauthorized sandwich.

Who knows what possessed him to do such a thing but Mr. Young decided to smuggle a sandwich into his suit and then on board the space craft. At the appropriate time he retrieved the sandwich from his suit and took a bite. He then offered some to a fellow astronaut. He quickly put the sandwich back into his suit but by then it was too late. Crumbs from the sandwich started floating around the space craft.

Now astronauts knew about the difficulties of eating and drinking in a zero gravity environment but somehow Mr. Young had a momentary loss of memory. Sure enough, just as was demonstrated by NASA many times before, crumbs from the sandwich started to float around inside the capsule and soon threatened to become a serious problem. The crumbs could have easily clogged the capsule's sensitive high tech equipment. According to NASA that was a text book "no-no" as every astronaut was well aware.

It was also exactly the reason why NASA coated all of the astronaut's food with a special material that would prevent that kind of thing. Not only was NASA upset but so was the US Congress. Law makers were irritated over the fact that tax payers spent "all that money", apparently millions of dollars developing special foods that were made to be safely consumed in space and not for a self-proclaimed "Homer Simpson" to bring his own food and threaten to create irreversible problems. That's right, the guilty party compared himself to Homer Simpson.

NASA thought it was a no brainier and that everyone knew that brown bag lunches were prohibited on space flights. In case you are wondering, the sandwich was corned beef and its taste was not out of this world, because taste buds in space are dulled to the point where fine chocolate can taste like candle wax.

By now we should know that spending time in space is much more problematic than NASA might acknowledge. What is assembled next is a story that could illustrate the point. When mother nature comes calling she does not care where you are, especially if you are in the cramped confines of space ship hundreds of miles above Earth. And once again astronaut John Young takes center stage.

You already know about Mr. Young from his corn beef sandwich that was smuggled on board his space craft but did you also know that he had a case of bad gas, as if there's ever any other kind, and shared his flatulence with much of the world. Here's what happened, astronaut John Young, was standing on the moon with mission control and much of the world either listening or watching when he developed an inopportune case of flatulence, the farts, the butt burps of whatever else you want to call it. Making matters worse, or more comical, depending on your point of view, he left his microphone on, and remember who was listening. Apparently the "hot mic" was not sensitive enough to pick up the unmistakable sounds of gas being expelled from John Young's body but it did record in embarrassing clarity the verbal comments that accompanied his flatulence. You can use your imagination, but many of the words used to describe his condition contained four letters and some were even used in colorful combinations. Evidently Mr. Young was without stage fright as he expelled the gas from his body with the same vigor that he expelled those colorful adjectives.

Many of us have had those unfortunate moments while in an elevator, or in a car and we do all that we can to remain composed and release nothing, sometimes going to considerable lengths to maintain containment. But apparently Mr. Young didn't know, didn't care or just could not help himself and put on a show that were are still talking about. It was a permanent recording of one of the more memorial, and smelly moments in the history of space exploration.

Here's where we go from bad to worse. The governor of Florida was so offended that Mr. Young blamed his flatulence on the state's oranges that the governor embarked on a media campaign to reassure the public that Florida oranges do not cause uncontrollable fits of flatulence.

Fact of the matter is, crew members on previous flights developed minor digestive problems. So NASA doctors increased the amount of potassium in the diets of crew members as a remedy. Unknown to astronauts at the time were the aforementioned side effects of uncontrollable gas presented by potassium. But it seems NASA was well aware of the potentially embarrassing problems of farts and weighed them against the more serious problems of how the body's digestive tract can be effected by the internal buildup of gas in a zero gravity environment.

On previous trips astronauts had trouble belching and instead pushed the gas into their intestines. Not wanting to find out what the result of excessive intestinal gas build up would do to its astronauts in space, NASA took the precaution to increase potassium levels in their diets, and did not consider the offensive odors or the embarrassing moments that would likely follow. After the story made headlines the word ass in potassium might have become more noticeable to readers.

Stuck on Earth!

Space has been called the Final Frontier but is it really the Ultimate Barrier? Consider the difficulties of human space travel which were initially exposed during those first efforts at traveling to the moon. As impressive as those accomplishments were the underlying lessons from those early space missions could be we really can't travel very far in the solar system.

When the vastness of the universe is taken into account, traveling to the Moon, our nearest neighbor at a distance of 238,000 miles, might be compared to a child expected to travel great distances without the necessary maturity or required experience. Consider our closest planet, Mars about 33.9 million miles away, could be too far for a human voyage to be worthwhile considering the significant amount of knowledge we've already gained from studying the planet for about 40 years with probes and space telescopes. Even though it might not be practical to travel to Mars at least consider the logistics. Human travel to the Red Planet would have to be planned while the orbits of Earth and Mars are nearest each other which only happens every two years and at best might only provide a window of a couple of weeks for humans to stay on the planet. Any longer than that and the orbits of the two planets would have them too far apart to make the trip back home realistic. In other words, if the timing is wrong the astronauts might die in space or perish on Mars.

Also the great distance of the trip would probably require two space crafts, one for supplies and the other for passengers. The price of such a trip would be cost prohibitive considering it costs about $10,000 dollars to send a pound of cargo into orbit. And that's not to mention NASA's budget has been dwindling for years, so much so that we can't even return our own astronauts from the International Space Station because the Shuttle Program was canceled after it was deemed too expensive. So that could be it! Are the Moon and possibly Mars are all we can expect from space travel? Are we trapped on Earth! Are we restricted by nature and limited by our technology to leave the planet?

Those are some of the questions posed by theorists who also ask, is our limited technology, the vastness of space and the short life spans of humans a natural design to keep people in check so as not to venture out into the universe? Is our natural character, described as violent and non-tolerant, a reason why aliens have not contacted us and helped us with their superior technology? In our present state would aliens see us as a liability and not an asset? A more radical theory posits we have been watched and judged and a verdict has been rendered: intelligent alien life decided not to get involved with the human race and instead chose to leave us to our own self destructive devices trapped on here on Earth.

Space is a deadly environment that in no way can be applied to the concept of Manifest Destiny where humans traveled across the North American plains in a hospitable environment. During the era of Manifest Destiny humans could breathe fresh air, drink clean water and eat the animals they killed.

It seems we have been humbled by the dynamics of space. Arguably we don't belong there. Humans will be killed the very instant that they come into contact with the lethal vacuum of space. Certain death awaits humans even at what could be called the pre-space altitude of 120,000 feet in the Stratosphere. There is no pressure, no air and no means by which human life can exist on its own.

As soon as we enter the weightlessness of space our bodies begin to deteriorate, the loss of bone marrow and muscle mass continue until we return to Earth. With no air to breath, we would suffocate but probably not before we were killed by a number of other deadly effects of zero pressure and no gravity. And there are the extreme temperatures in space, lows reach minus 457 degrees below zero and the highs are well into the millions of degrees, far beyond the 2,200 Fahrenheit required to melt rock.

The old saying "If man were meant to fly he would have wings" was rendered mute by the invention of the airplane. It seems there is no such invention that can disprove that saying as it applies to space. So what can we do in the interim while waiting for technology to advance to the point where humans can travel significant distances in space? How about returning to the moon! We don't have many options. Work with probes and telescopes continue to yield excellent results and are said to be well worth the cost but they seem to be unable to solve the problem of human space travel.

During the early stages of space travel humans could have been too preoccupied with looking past the moon and to traveling to other planets. Was this approach premature? When we looked past the moon, did we literally look past the opportunities and lessons the moon could have taught us about space travel? If the galaxy is seen as the neighborhood and the moon as the sand box, should we not play in the sand box until we are old enough to venture out into the neighborhood. Is there not more we can still learn from trips to the moon? One major objective of any space program is to discover intelligent alien life. Are we more prone to discover alien life on Earth or out in space while conducting exercises and experiments traveling to and from the Moon?

Could the moon be home to an international team working on its surface? That was one of the objectives of the Constellation Program which will be reviewed later. Consider the benefits, the more we work together with other nations the less likely we are to go to war against them. The moon could also be home to a lunar telescope that could save lives on Earth and here's how. The telescope could cover our "blind side" when it comes to looking for asteroids and comets that could wreak havoc on the planet. Remember the asteroid that broke up in the sky over Russia in February of 2013! That asteroid gave no warning of its impact because it came from the direction of the sun. We've searched much of the sky and have documented 13,279 asteroids and comets that could impact Earth. But telescopes on Earth cannot look into the sun and that is the direction from where the 2013 Asteroid came. In other words, it came from our blind side. A lunar telescope could provide a lifesaving remedy to the blind side problem. A growing consensus in the scientific community seems to reassess the goals of our space exploration programs that would better fit our strengths and are modified to our weaknesses.

Which Way is Up?

From the moment we begin life to the moment we die we're in a constant battle against gravity.

The battle is so intense we don't even realize we are engaged in one. Gravity is relentless, beginning when a baby struggles to stand and take those first steps to when the elder is so aged that he or she can no longer stand under his or her own weight. We accept gravity as a part of the natural world. But a closer look might raise more questions than answers.

The strength of gravity is different throughout the universe. A 180-pound human on Earth would weigh about 30 pounds on the moon because the moon has about 17% of the Earth's gravity. Jupiter has about twice the gravitational strength of Earth so that same 180-pound human would weigh about 360 pounds on Jupiter. As strong as gravity is on Earth it is much stronger elsewhere. Imagine a planet or star with a density and gravity so great that a tea cup of its material would weigh more than a mountain! Imagine an Earth sized planet or Star compressed into an area the size of Chicago, a grain of its sand might weigh hundreds of tons. Scientists say such phenomena has been observed as we learned with neutron stars.

Like so many discoveries, part of gravity's secret was unlocked by serendipity or in other words by accident. While in orbit an American astronaut was looking at rice or something similar in a plastic bag. Remember he was in the weightlessness of space. He noticed that the rice pieces would naturally clump together. Apparently the pieces were drawn toward each other by their own gravity. When the astronaut would press the plastic bag between his two fingers, the bits of rice would separate. But once he stopped pressing the plastic bag with his fingers, the bits of rice would collect back together into one lump. Gotta love serendipity! Long story short, the astronaut relayed the new found information to astrophysicists and other scientists who further studied the phenomenon. They found that if a handful or rocks were thrown on the ground they would just stay there, held in place by Earth's gravity. But if that same handful of rocks was thrown in space, the rocks would form a disc and in the weightlessness of space those rocks would be drawn together, attracted to one another by their own gravity and again form a larger clump just like the bits or rice.

Hence smaller objects in space such as rocks and asteroids and drawn to larger objects by the universal law of gravity. When this natural process repeats itself enough times the end result is really a large object like an asteroid that continues to grow by consuming smaller objects. Eventually the asteroids become so large that their shapes change. The original shape may be similar to a potato or a dog bone. But once they are big enough, they become perfectly round as in a sphere, which explains the shapes of large moons, asteroids, planets and stars. This is done by the object's own center of gravity. Scientists argue that once asteroids reach about 500 miles in diameter its gravity becomes strong enough that it will force its shape into a sphere. If it's smaller than 500 miles in diameter its gravity is too weak to overcome its shape and it will remain configured like a potato or a dog bone. Gravity is strongest where the mass is the greatest, such as the center of the equator on a planet. This can explain why a moon would orbit at the center of a planet and not at its polls. So once something has enough mass its increased gravity will force its shape to become a sphere.

All objects exert gravity. A person could walk by a wall and that wall would exert its gravity on the person. Of course the person would not feel the wall's gravity because Earth's gravity is so much stronger.

When you understand gravity's true nature you realize that there really is no up or down. There is only the pressure exerted by gravity. As one scientist put it, "The apple did not fall to Earth, instead the apple collided with Earth," which begs the question, which way is up? In space the answer is "up" is any direction you choose and the same holds true for "down."

Fire without Air!

Ever wonder how there can be fire without air! I first wondered that years ago when I saw one of NASA's pictures of a solar flare on the sun's surface. The flame, which can be described as heat, light and radiation was three times the size of Earth. Once I got past that amazement I thought how can that be, fire in space without air?

Apparently there are at least two kinds of fire, the one we see here on Earth and the other that comes from the Sun. Many scientists might disagree with this description of fire in space but after reading material from many scientists who describe the sun as a fiery place I stand by that description and that's not to mention the many pictures released by NASA that show huge flames on the sun's surface during its many periods of extreme and violent activity.

Fire on Earth needs air for fuel, not so on the Sun. Fire on Earth is a chemical reaction, not so on the Sun. Fire on Earth involves combustion, again not so on the Sun. Fire on the sun is a nuclear reaction that involves fusion instead of combustion. Earthly fires consume material through combustion that chemically changes that material forever by burning it to ash. Fire on the sun is the result of massive pressure and extreme heat that force atoms together known as fusion. When two atoms are pressed together with enough pressure they release heat, light and radiation that we see as fire on the sun. If humans could ever produce this kind of fire we might create an endless supply of energy, not to mention we just might recreate photosynthesis. Photosynthesis is the process by which plants produce life from sunlight, something artificial light cannot do. Think of the advantages of sun light produced by humans, the possibilities might be endless. We could produce enough food and energy for them never to be a problem again. With those fundamental problems solved humans might have enough time to focus on solving other problems.

The sun's nuclear fusion begins inside its core. The temperature there is about 23 million degrees Fahrenheit. The pressure at the Sun's core is about 340 billion times what it is on Earth. So what does the Sun have that the Earth does not? Gravity. We already know it takes two hydrogen atoms and one oxygen atom to create water. We also know that the combination of four hydrogen atoms can be used to produce an explosion powerful enough to destroy islands. Scientists are now using high powered lasers to discover new methods of increasing pressure which might hold some of the clues in creating fusion.

Another amazing fact about the sun is the farther away you travel from it, the hotter it gets. The sun's surface temperature is around 10 to 15 thousand degrees. But travel about a million miles away and the temperature doesn't drop, instead it goes significantly up, to around two or three million degrees. How can that be? The farther you travel from a fire on Earth, the less heat it emits. But a million miles from the Sun and the temperature sky rockets.

This area of the sun's atmosphere is called the Corona. Here's a possible explanation for this phenomenon. We know that heat rises. The hottest part of a flame on Earth is just above tip of the fire, again because heat rises. The same dynamics might apply to the sun. But remember in space there is no up or down only the force excreted by gravity. So as heat escapes from the sun it dissipates at an even rate that surrounds the entire star. Just like the flame on Earth that gets hottest at its top, the same applies to the Sun but in this situation the rising heat surrounds the entire Sun equally on all of its sides and rises from those sides at the same rate which can explain why its temperatures are hotter the farther away you get from its surface.

Back to fusion and the flame that needs no air. Before 1950 it was questionable if we even knew how nuclear fusion worked. Today we have a fairly good understanding of the process. Again scientists are now using super powerful lasers to shoot a beam into a small, confined, funnel shaped area in an attempt to create the kind of pressure required for sun like fusion. While humans still cannot replicate the kind of pressure we have produced temperatures that exceed those inside the core of the sun. If creating sun like fusion only requires massive pressure and extreme heat, then could we say we are half way there? If we have come this far since the 1950's what will the next 50 years hold?

But Why Pluto?

Another unique personality at the forefront of astrophysics is a man whose character is as colorful as the aforementioned mastermind behind Rosetta. This man also came from modest beginnings, he was a product of the public school system in a large inner city. He too is a devoted father, husband and family man. And he too, like the man behind Rosetta is not only blessed with impressive intelligence, but also has the ability to impart that knowledge in understandable terms. But this man received hate mail from many angry children in elementary school because of what he did. He helped demote Pluto from the ranks of a planet.

Any number of scientific publications such as NOVA carried stories about hate mail from third graders to the man who helped remove Pluto from its planet status. One child wrote "I do not like your anser!!!Pluto is my faveret planet. You are going to have to take all of the books away and change them." Another student wrote "Do people live on Pluto? If there are people who live there they won't exist... Please write back, but not in cursive because I can't read in cursive." His response: "I got blamed for killing Pluto, but I was just an accessory to it. I drove the getaway car for it... Pluto, I am blamed for it, but I want to alert you that it is still not a planet."

But why would someone demote Pluto? The answer might lie in the method behind what the angry school children call the madness. The criteria established to remove Pluto from its long held planet status seems to be three fold. The first, to be a planet Pluto must orbit the Sun, which Pluto does. The second, Pluto's mass must be large enough to form a sphere, which Pluto's is. But the third and final criterion is Pluto must dominate its own orbit, which Pluto does not.

If Pluto had a smaller moon Pluto might still be a planet. Pluto's moon is called Charon and Charon is almost the same size as Pluto which means they both have the almost same gravitational pull on each other which makes it impossible for Pluto to dominate its orbit. Instead they both kind of revolve around each other as they go around the sun together. Conversely, Earth dominates its orbit as the Moon submissively orbits Earth as Earth and the Moon both go around the Sun.

Another possible reason for Pluto's demotion might lie in the general composition of the solar system.

The solar system consists of our sun at the center with eight planets. There used to be nine. Half of them are gas and half of them are rocky. The inner planets, Mercury, Venus, Earth and Mars all have hard rocky outer shells. The outer planets, Jupiter, Saturn, Uranus and Neptune are all composed of gas. Pluto might have stood a better chance of retaining its planet status had it been closer to the Sun where all the rocky planets are located.

At the outer limits of our solar system near Neptune and Pluto is something called the Kuiper belt. The Kuiper Belt acts like a perimeter around the entire solar system and it is made of asteroids. Some asteroids are big enough to be spheres. While Pluto is a sphere it is still small for a planet. One deciding factor in Pluto's demotion could have been its proximity to the Kuiper Belt, which is full of asteroids and not planets. In other words, all of the rocky planets in our solar system are closest to the sun and not on the outer perimeter like Pluto.

In fact, the planet is so small that you could fit two Pluto's inside one United States. Pluto is much smaller than any other planet in the solar system and might be more similar to an asteroid than a planet.

So was Pluto guilty by association? Don't know! But we do know Pluto it is no longer a planet and it's not an asteroid, it is something called a Dwarf Planet. A dwarf planet is a body in space that resembles a small planet but does not meet some of the aforementioned criteria of a planet. The strange paradox for Neil deGrasseTyson, the man blamed for "killing Pluto" may not be fame for discovering a planet but instead infamy for killing one.

New Horizons!

The New Horizons space craft is yet another milestone for the space program, launched in 2006 it has traveled a long way, some 3 billion miles to get close to what is called an icy dwarf planet. An icy dwarf planet is an object too big to be an asteroid but too small to be a planet. This one is Pluto, downgraded from its planet status but still making news. When we say close we mean New Horizons came within 8,000 miles of Pluto in the summer of 2015, very close by galactic standards. As part of the mission statement the craft will send back photos and data to Earth which will certainly increase the already substantial knowledge base. This is the first time any space craft from Earth traveled to what used to be the ninth planet which is still called Pluto.

It was also one of the fastest space craft ever launched blasting along at more than 36,000 miles an hour, which means it could travel around Earth in about 45 minutes. It passed the moon in about nine hours. It took the first Apollo mission about three days to make it there. Still at 36,000 miles an hour it took New Horizons about nine years to travel to Pluto. When New Horizons took off Pluto was still a planet.

Because it is so far away solar panels are not of much use to New Horizons so it is powered with nuclear fuel, running on something called a radioisotope thermoelectric generator or RTG. The generated heat is used for power. It can put out a maximum of about 300 watts, which is the equivalent of an average kitchen blender in your home. Scientists have fun with this by saying things like, "is the public to believe that there is a probe racing through space for the last nine years at 26,000 miles an hour and is about 3 million miles away with a kitchen blender for an engine" funny, but actually true. As weak as the kitchen blender engine is in New Horizons it is still about three times as powerful as the motor in the rover Curiosity which is still at work on the Martian surface. New Horizons got most of its power during takeoff and it carries a special type of fuel that does not need to be ignited, instead it goes through a chemical reaction that produces a superheated gas.

This is one of those proverbial cases where slow and steady wins the race. We've already learned about the procedure called a gravity assist from the space craft Rosetta. Well the technique was used by New Horizons when it passed the giant planet Jupiter. The strong gravity from Jupiter's sling shot effect increased its speed by about 9,000 miles an hour. If not for that boost, the trip would have taken another three years.

Jupiter is known as a friend to Earth because Jupiter's gravity prevents many stray asteroids from crashing into our planet. As the probe passed the giant gas planet it recorded volcanic activity on one of Jupiter's moons. Even from that great distance it sent back a picture, a high resolution image to Earth which took about 45 minutes to receive. New Horizons continues to move past Pluto but its mission might not be over, not if NASA can find additional funding. The goal is for New Horizons to travel to the Kuiper Belt which acts as sort of a boundary line for our solar system. This belt is full of asteroids that can hold answers to some of the mysteries of the early universe and how it was formed. After some stiff competition the New Horizons effort is expected to put the United States back into the lead of the international space exploration race, with a kitchen blender engine doing all of the hard work.

Can't Explain Everything!

Many people believe that science can explain things that religion cannot which might be why those folks put more faith in science than in the clergy. People once thought the Earth was flat as far back as 550 BC and continued thinking so for more than a thousand years until they were enlightened by science. One definition of Science: the intellectual and practical activity encompassing the systematic study of the structure and behavior of the physical and natural world through objective observation and experiment.

But at the advanced level Science seems to contradict itself. Consider the Big Bang theory, it attempts to answer some of the most basic questions of human existence. Where did we come from? Where are we going? How was the universe created and why are we here? Here the Big Bang theory does little better than religion in providing answers. The Big Bang is the most commonly accepted scientific theory but it asks us to believe and awful lot with little proof. Religion does the same thing. Just think about the absurdity of the Big Bang theory, one moment there is nothing and then poof! Wa-la there's the universe. It seems we are to believe the most outlandish stories from our most intelligent people.

The Big Bang theory suggests something was created from nothing. Wow! Scientist argue that all this happened in a micro instant almost 14 billion years ago. One moment there was nothing and the next there was this massive, ever expanding universe, trillions of degrees hot at first and then cooling down to the point where elements could eventually form into life. One great thinker was quoted as saying:

"Follow the evidence, where ever it leads. If you have no evidence reserve judgment." That great thinker helped to demote Pluto and he has another quote: "The universe is under no obligation to make sense to you." Was the Big Bang the beginning of time? What happened before the Big Bang? Why doesn't the scientific community just collectively say we don't know? I don't know! But considering the short spans of our lives, we might never know these answers – still many people chose to be guided by reason and not faith and given enough time, thought and research science will eventually provide those answers.

Consider the sun, a common belief is that we owe everything to that star. So the obvious question is where did the sun come from and how was it created? We don't know, but just like the Big Bang there are well accepted theories put forth by brilliant and believable people. The most commonly accepted theory is the sun was created from the dust and gas in a cloud called a nebula that was mostly hydrogen and helium which was left over from the Big Bang. Problem is, it's just a theory. Common logic states it is plausible to be build a theory on a fact but don't build a theory on a theory, but seems that's what's been done here. The nebula theory is built right on top of the Big Bang theory with no factual foundation in between. Again many people chose to believe the sun was formed from a nebula cloud of gas and dust, at least until a more believable theory is presented.

Theories about the Big Bang and the formation of the sun seem to be devoid of sufficient evidence, facts and empirical research. Under different circumstances many people would in no way take such notions as serious. One definition of research is, the available body of facts or information indicating whether a belief or proposition is true or valid. What about a fact, which is defined as a thing that is indisputably true? A large majority of the topics in astronomy can be disputed. And finally, what about empirical research? Empirical research is based on, concerned with, or verifiable by direct observation rather than theory or pure logic. We can directly observe the moon because we've been there but that's about it. There is indirect observation through the telescope and most of the rest is theory.

Are there some things that we are not meant to know? Can we live with that? Do we have a choice? What if there were scientific, factual evidence based on empirical research about the origin of the universe and the sun? All we have is theory. Could the universe be kind of living environment which is not meant to be fully understood and sort of keeps us in check by our short life spans and its great inhospitable distances?

Welding without Heat!

Cold welding is an amazing fact in space whose names explains its process. Simply put, metal sticks together in space. If two pieces of identical metal touch each other they will permanently bond together in the vacuum of outer space, just as if they were welded together on Earth but without a welder's torch or any kind of heat. This was first discovered in the 1940s. Scientists put two flat metal surfaces together in a vacuum and found that they bonded together so tightly as to replicate a wield. All of the atoms in the experiment are of the same kind, so as strange as it sounds the atoms have no way to know that they are different and as a result the stick together in one large clump. This would not happen if some atoms were copper and others were aluminum.

The process takes place with no heat, no liquid and with no fusion. That doesn't happen on Earth because we don't live in a vacuum. On Earth the atoms in water and in the air create separation but in space similar atoms in metal stick to each other and it's a problem for space craft. Consider what might happen between two moving metal parts of a space craft that came into contact with each other. Or consider what could happen if two space craft made of the same metal compounds touched. It is a phenomenon that engineers take into account when designing space craft.

The process of cold welding might have caused problems with the Galileo satellite which resulted in some unexplainable anomalies that baffled scientists. Cold welding also has many engineers altering the compounds used to construct space ships and inserting dissimilar metals so as to avoid this unwanted bonding process. Where metal selection is not an option, engineers have been forced to coat similar metals with substances that will not attract each other in the space. Thermally insulated parts that would reduce interstellar exposure are yet another possible remedy to the cold welding problem.

After understanding the dynamics of cold welding, many people wonder do we even belong in space? Consider the endless vacuum of space, thought to be a constant sucking force that would quickly siphon our eyes from our skull because there is insufficient pressure to hold liquids together. This endless vacuum that can slam similar metals together with such force as to weld them together permanently repeatedly begs the question, what are we doing there? It's clear our bodies were not equipped to deal with outer space. On Earth humans grew bigger brains and evolved to walk upright in order to adapt to our environment. Space is nowhere near comparable. Are the dynamics of the cold weld just another harsh reminder of our limitations in space? The environment of space constantly demonstrates that much of its composition is just too mysterious and too deadly for human habitation? Conventional wisdom would seem to suggest to that space is no place for humans and the bizarre process of cold welding seems to be another example of this idea. But there are many people who say there could be worse things than being devoid of conventional wisdom.

The Eternal Moment!

While on the topic of the vacuum of space, guess how long the foot prints of the Apollo astronauts will last on the moon? A lot longer than they would last on Earth. The moon has no atmosphere, no air, no water, nor wind and erosion to change its surface. This is why craters from asteroid impacts remain virtually unchanged and visible on the moon for hundreds of thousands or even millions of years. Again this does not happen on Earth. That's because our planet's atmospheric dynamics erase and erode marks left on its surface. So how long will those moon footprints last? An estimated 100 million years or longer. But while footprints might last a long time that doesn't mean they'll remain unchanged.

Consider the flags on the moon. There were six American flags planted on the solar surface in the 1960s. All but five are still standing. But those American flags are no longer red, white and blue. They're all bright white, bleached that way because of the sun's ultra violet radiation is unfiltered by the fact that the moon has no atmosphere. In theory that radiation could disintegrate the flags long before the foots prints fade away.

As for what happened to that 6th flag no longer standing, well stuff happens, even in space. It was blown over by the thrust of the Apollo 11 craft. Astronaut Buzz Aldrin remembered looking out of the window and seeing it happen. Even though the United States was not the first nation to reach the moon, the Soviets beat us there in 1959 with an unmanned craft; America can boast it is the only nation to have put its citizens on the lunar surface, 12 of them since the first astronaut uttered that "one small step for man..." declaration in 1969.

As you may or may not know the Moon does not spin on its axis as does the Earth which means the Moon always shows us its same side. But this does not mean that one side of the Moon sees more Sun than the other. Both sides of the Moon see equal amounts of Sun light because of its synchronous rotation. Did you also know "The Dark side of the Moon" is not an accurate term? The term should be the far side of the moon. Just a little more debunking. The moon is not round or spherical. It is shaped like an egg with one of its small ends always pointing right at us. This explains why the egg shaped moon looks round. And the moon is also moving away but it is in no hurry to leave. Just like a ball at the end of a rope that is swung around in a big circle the Moon is moving farther away from Earth at about an inch and a half per year. But too much debunking can be bad for the imagination, the dark side of the moon sounds so much better than the far side.

How's your Hearing?

We are told that space is silent, apparently all of it is completely without sound. But not everyone agrees. If space really is without sound that could be a good thing because we are told if we could hear the sun's surface activity it would be so loud we would become deaf. Sound waves need some sort of medium to travel through and off of which they can vibrate or resonate. In many cases the denser the medium, the faster sound travels. Of course sound travels at different speeds through different materials depending on their densities. Sound has been described as a pressure wave and the greater the pressure, the greater the speed. Sound traveling though Earth's atmosphere moves around 760 miles an hour. But there's no pressure in the vacuum of space. When astronauts are inside the space craft there are no communication problems, at least none that came be blamed on a vacuum. Air is in inside a space craft so he or she has no problem hearing their voice and talking to others. Even outside the ship they can hear themselves talk, if they are wearing a space suit because the space suit contains air.

But how do astronauts speak with each other when there is no atmosphere, air or sound waves?

They communicate with radio waves. But NASA is not so quick to sign off on this notion of a soundless space. NASA researchers argue there are gasses in space that under certain conditions can behave like sound waves on Earth. The difference is these interstellar gas clouds are much less dense than the ones on Earth. So if sound in space traveled through one of these big interstellar gas clouds, just a few atoms per second might travel to the human ear and that frequency would be too low for humans to detect. But NASA states that it does have sensitive equipment that can actually hear in space. It has designed special equipment that can record electromagnetic vibrations and then transfer those vibrations into sounds that are detectable by the human ear.

Why do we care if we can hear or not in space? What benefit does this confer to our daily lives on Earth? Answer for most of us is probably nothing, except for the fact that these sorts of celestial matters seem to add an element of quality and interesting entertainment into to our lives. If nothing else, they punctuate the boredom that can come with a routine daily existence. Of course it's nice to imagine a world so different than ours that is not dominated by honking horns, screeching tires and screaming irate humans, crying babies, demanding bosses and so one and so forth. You get the idea. Well imagine, if just for a moment that all of it went silent like space. Peaceful isn't it. And therein may lie the value of learning about the silence of space, just waiting to remind you of its worth every time the earthly sounds of everyday life become a bit much.

The Big Rock with lots of Potential!

Apparently there is an asteroid so big there is some debate if should be classified as a dwarf planet, which makes its situation similar to Pluto's, the smallest planet in the solar system demoted to a lower status. This big space rock is called Ceres. It is some 600 miles wide which is about 100 miles larger than the size required for a space body to become a sphere, the result of its gravity transforming its shape into something perfectly round.

The surface area of Ceres is roughly equal to that of Argentina and it resides in the Asteroid Belt which is the area between Jupiter and Mars. If Earth gets hits by an asteroid, it will probably come from the Asteroid Belt which makes this area and its biggest asteroid worth taking a closer look. Ceres was the first asteroid discovered in 1801. At first it was called a planet, kind of like Pluto. Like many objects in space it was named after a deity but more important it might have water, which according to scientists could indicate life on the planet.

The water in Ceres might be contained in an underground ocean and it also has an atmosphere which increases the potential for life. Problem is, Ceres is a long way from the Sun, maybe too far for life as we know it to evolve. Scientists are optimistic that despite its intense cold, life might have taken hold on the planet where surface temperatures could rise to 37 degrees below zero, primitive bacteria on Earth survive in similar conditions.

So how many asteroids reside in the asteroid belt and what are the chances one of them will hit Earth? According to the best calculations there are about 600 asteroids in the belt that are larger than 60 miles.

Ceres is among the four largest asteroids in this belt and these four make up about half of the material in that belt. According to come calculations, your chances of getting killed by an asteroid are about 1 in 700,000. That's about the same chance as drowning in a flood or getting killed in a fireworks accident. The latter possibility was publicized in the summer of 2015 after a star NFL player with the New York Giants lost part of his hand in a fireworks accident. So it can happen. Of course the chances of 1 in 700,000 don't rule out the possibility that you could get hit by an asteroid before you finish working this week. Remember the asteroid that seemingly came out of nowhere and exploded in the sky over Russia on February 13, 2013. And that's not to mention some of the newly discovered asteroids that are so dim they can barely be seen until they are almost on top of us. So asteroids are the topic of many discussions: when and where they might hit, if they carry water and are they valuable, that's right a profit might be gained from asteroids.

So what's an asteroid worth? That depends on whom you ask but chances are they'll be worth more in 50 or 60 years. On Earth as we continue to deplete our natural resources there will come a time when there is greater motivation to look elsewhere for replacements. Here's where we insert the possibility of asteroid mining. This is not science fiction but it's not quite science fact, not yet. There is speculation that some of Earth's most valuable elements could be exhausted in 50 or 60 years. These include elements like phosphorus, antimony, zinc, tin, lead, indium and gold and copper. Our technology is not quite there yet but the idea is we could basically mine asteroids, of course were talking about the ones that come close enough to Earth to make this possible. Water could also be extracted and besides human consumption it could be mixed with hydrogen to produce fuel. As we'll find out later just one asteroid could contain more fresh water than all of Earth, which only adds to their potential value. Think of what technology made possible 50 years ago and then think of what might be possible in another 50 years.

Advocates for asteroid mining point out that it's simple for a scientist to make rocket fuel from water and in space the separated hydrogen and water can be used to refuel craft at space stations. This would provide an enormous advantage because space craft could then travel farther and longer without having to rely on Earth as a fuel source. One scientist compared to a car that can't travel very far with only one gas station located in the hometown of the owner. Some of the best places to mine natural materials on Earth are where asteroids have already struck which provides motivation to get the materials directly from the source where they are more abundant. In one sense these new space prospectors are just like the prospectors from the California Gold Rush in the mid-1850s in that they survey the best rocks with the most potential but that's where the comparison ends. Space prospectors would have to deal with the lack of gravity, space and time, extreme temperatures, costs and the effects of prolonged exposure to the human body to name a few, all of which were not obstacles for the prospectors of an era gone bye.

Like many other things, asteroids vary in speed, shape, size and distances traveled. Some are as slow as 25,000 miles an hour, if you could call that sluggish. Other asteroids are as fast as 160,000 miles an hour. When it comes to fast moving space rocks, speed can make up for size, relatively small ones traveling very fast can cause surprising amounts of damage upon impact, sometimes more than the slower, bigger ones. The smaller ones could fit in your hand and the largest is the size of Argentina.

What about meteors? I don't know why scientist do this, but meteors are just smaller asteroids and if that's not confusing enough, they came up with three names for the same object. When it is traveling in space it is called a meteoroid. When it is traveling through Earth's atmosphere is it called a meteor and when it has landed on Earth's surface it is called a meteorite.

Never a Good Time for a Rouge Planet!

Some of the theories about rouge planets are amazing including one idea that suggests a rouge planet could destroy Earth without impacting it. One of the more popular rogue planets is called Planet X, known in ancient times as Nibiru the Destroyer or The Terrible Comet. A rouge planet is a planet that does not have a star to orbit and instead wonders around the universe unnoticed until it disturbs or destroys something.

One theory is a rouge planet enters the solar system once every 3,600 years or so. During this time, they may pass by the Asteroid Belt and dislodge a few asteroids as a result of their gravity. One or more of these asteroids could be flung out of the solar system or propelled in the other direction toward Earth. If the asteroid hits Earth the rouge planet would be to blame even though it never crashed into our planet.

Another possibility is the rouge planet passes by Jupiter and is effected by the strong gravity of the massive planet. Jupiter's gravity is so strong it is seen as a protector of Earth by securing in place many asteroids that otherwise might career into our planet. Jupiter is also the protector against long period comets. Just like its name sounds, these are long term objects that travel far out of the solar system on orbits that cover huge distances and take lots of time. An example came in 1994 when the Comet, Shoemaker- Levy crashed into Jupiter. The event was recorded and then used to solicit money from congress to fund programs to identify similar objects headed Earth's way. Had Jupiter not been there, scientists are not quite sure what would have happened but they are sure had those same comets impacted the Earth it would have destroyed our planet. Put another way we all could have been killed years ago if that comet had a different trajectory that brought it into Earth.

As the rouge planet travels past Jupiter it's trajectory or path is changed to the point where it is now headed toward Earth. While it does not contact Earth it does come close enough so as to change Earth's orbit around the Sun. Earth's orbit is nearly circular but because of the gravitational change caused by the rouge planet Earth's orbit is now elliptical. At first there is nothing more than a subtle change.

At its nearest approach toward the Sun, Earth is much closer than before and at its farthest approach Earth is more distant from the Sun than it used to be. The effects soon become significantly noticeable. Earth is no longer in the "Goldilocks Zone" where its environment is not too cold and not too hot but just right. The summers on Earth reach around 140 degrees, just warm enough where the human body can no longer perspire fast enough to cool itself off. Heat exhaustion will force the body to fail. In the winter Earth is too far away for human habitation. The ice on the polls continues to spread. The increased ice continues to force more heat and sunlight back into space and the cooling process starts to feed on itself. Earth is now facing dire straits, its prognosis is not good and none of it was caused by a direct impact. Instead our demise was caused by a subtle disruption of the delicate balance of Earth's position in the solar system, again brought about by a rouge planet that never touched our planet.

Flat Expansion or Round Contraction?

Is the universe flat or round? Remember there was a time when humans thought the Earth was flat, they were convinced. Could we make the same mistake again? One of the currently accepted theories is the universe is flat. One scientist says the background of microwaves and the distribution of galaxies confirms the theory. The microwave background is basically heat and radiation left over from the Big Bang and it appears the microwave background and galaxies are laid out in a linear fashion. In a flat universe, parallel lines would never meet, instead they would run side by side until they reach some sort of end. They also say this formation was created immediately after the Big Bang.

Consider the dynamics of gravity. As mentioned earlier, a hand full of rocks thrown out in the space will form a flat disc and begin to rotate in a closing circular fashion, moving closer to each other until they are one large clump, drawn together by the individual gravity of the rocks. Is this a universal law? Can it be applied to the theory of a flat universe? What's under the universe if it is flat? What's above it? Problem with theories is they can raise more questions than they answer. So what would Einstein say? He already said something like matter and energy curve space time so he doesn't appear to be an advocate of the flat universe theory.

Consider another option. The universe is actually curved or even a round sphere as are the planets and stars. If the universe is a sphere, is there no beginning or ending but instead just a circular concept? Is a curved or spherical universe so big that it appears to be flat, just as it would to a person standing on the plains and not being able to see Earth's curvature which is obscured by its horizon line. Put an insect on top the largest hot air balloon and the insect might not be able to perceive the balloon's curvature; would it not then think the balloon is flat?

What about the notions that the universe is expanding and infinite! Two more theories. Consider the latter first. How can something have a beginning but no ending? If it has no beginning does it even exist? If it is infinite could the universe be round like a sphere? Could it sort of wrap around itself?

This might explain that the universe has no end, because of its round or cyclical formation, but it does not explain the question of its beginning. Problem might be we can't step back from the metaphorical trees to see the forest. What about expansion? A commonly accepted theory is that some 13.8 billion years ago, immediately after the Big Bang the universe expanded from something very small to something so big we cannot understand its scope, in part because it is moving outward at the speed of light and has been doing so for a long time.

In 1998 the Hubble space telescope identified far off supernovas, some are big stars in the process of exploding. The Hubble work suggests that the universe is not only expanding but its expansion rate is accelerating. It argues that the driving force behind this is something called dark matter. Dark matter has been described as material that does not reflect light and can take several shapes and forms. We are told this matter makes up about 27 percent of the material in the universe and you guessed it, it can't be seen nor explained. The clergy must be having a good laugh right about now. Reason being they are often criticized because they are guided by faith in the absence of fact, scientists seems just as guilty, believing in these theories in the absence of fact. Is there such a thing as a scientific leap of faith?

Wasted Energy!

Too bad we don't know how to do a better job of harnessing the energy from the sun. If we were more competent we might be able to solve many problems on Earth. According to scientists the Sun showers Earth with more energy in 14 seconds than all of the people on the planet can use in one day. There are ample amounts of excess energy coming to us from space that can be put to good use on Earth. Could part of the problem be that the authorities of established energy sources feel threatened by innovation and competition? Think about all of those corporations profiting from fossil fuels, coal, electricity and the like. Just think if they were advocates of power from the sun, proponents of solar power.

Solar radiation is absorbed over most of the globe including its land and oceans with water making up about 71 percent of the planet. Every hour, for billions of years without disruption Earth's atmosphere has absorbed huge amounts of the Sun's energy, which is another tremendous resource that continues to be ignored. That's more energy from the Sun in one hour than the Earth uses all year. In one year so much solar energy bathes Earth that it adds up to more energy than we could ever produce, using coal, oil, natural gas and compounds extracted from other mining operations. If the Sun can convert 700 million tons of hydrogen into almost as much helium every second, what else can it do? And if loses 5 million tons of material per second, how much energy must it generate every second in order to maintain its form? And again why can't humans find a way to use some of that lost energy?

Assuming at some point humans will acquire the ability for significant space travel we might be able to leave the solar system and travel out in the Milky Way where there are some 100 billion stars. If half of them are similar to the sun than there might be 50 billion additional sources of fuel. If we chose to reduce our ambitious projections then we can consider there are some 6000 stars visible to the naked eye, which poses the possibility of at least 3000 options for fuel and energy capable of supporting humans for extended periods in space.

Ponder the fact that the sun puts out about 386 billion billion megawatts of energy. One megawatt hour is worth about 150 dollars of electricity. Think how many appliances and heating systems are powered by 30 dollars in one month's electric bill. The sun's energy has been around and will continue to be here for a long time. It's been producing energy for about 5 billion years. It can take millions years for the energy produced inside its core to reach Earth and if the Sun were to stop producing energy today, it would take millions of years for the effects to be felt on Earth. It is a powerful, stable source of energy that has tremendous longevity.

The sun is mostly hydrogen and helium, some 90 percent of the former and some 8 percent of the latter so why not begin exploring new and more efficient ways to use and exploit hydrogen and helium. The same approach could be applied to neutrons and protons. Since the Sun also produces these elements in abundance why don't we on Earth find new and better uses of these elements? If you could take a small area of the sun's surface, something about the size of a quarter, that quarter would shine as bright as 1.5 million candles. And dark spots on the sun are not really dark, in fact they are about as bright as a full moon, but with the sun's blinding brightness as a backdrop the dark spots seem black. These dark spots generate tremendous magnetic forces from which we can learn and benefit.

The total amount of energy from fossil fuels used by all of humanity since the beginning of recorded time is less than one month's solar energy emitted from the Sun and showered on Earth. So what better time than the present to start utilizing its energy in a more efficient manner. The sun's potential is tremendous and can do so many things in addition to producing energy. But possibly the best thing about all of that energy that we use and all of the greater amounts that we don't, is it's abundant, it's nearly permanent and it's completely free.

Mirror, Mirror on the Wall...

Possibly the most beautiful spectacle in all of the universe is the comet, often referred to as a shooting star. They have had our attention and have been the object of our admiration ever since we saw the first one. Comets differ from asteroids in their basic composition and their trajectory. Comets are made up of dust, ice, rock and organic compounds. Asteroids usually lack the latter and it's this organic compound in comets that so excites scientists. There is so much excitement that as we mentioned earlier, it prompted scientist to land a craft on a comet to see if some of those organic compounds might contain amino acids. Amino acids are thought to contain the building blocks of life. Comets are also thought to contain water in large amounts and might have brought huge quantities to Earth to either fill or supplement our oceans.

A comet's path through the solar system is usually elliptical. Asteroid's usually orbit in a cyclical manner. Comets have tails produced by the gasses they contain. This is absent in asteroids. Some of the tails of comets linger in space for a million miles. Its tail is thought to be made up of the least substantive material in the universe. The tail material is extremely fragile, so scientist tried to capture some of it in what amounted to a net made of toilet paper. They were successful but when the material from the comet's tail was brought back to a laboratory on Earth the material was so delicate that it just fell apart, basically disintegrating right there in the lab.

Back to the comet's elliptical orbit. Its path takes a comet close to the sun at its nearest approach and then a great distant from the sun at its farthest orbit. Short period comets orbit the sun every 20 years or so. Long distance ones come by every 200 years or longer. Comets also have an atmosphere called a Coma. Its tail and Coma are seen when sunlight shines on them. As a general rule, comets might be smaller than asteroids. There are an estimated 5,200 comets in our solar system, there could be more than 150 million asteroids in the same area. Comets and asteroids also have much in common, including both crash into Earth.

Comets can also have infamous reputations. No asteroid has the dubious title of being involved with a mass suicide, as least none in the modern era. Remember the group called Heaven's Gate? It was based in San Diego, CA in the 1990s. In 1997 police found the bodies of some 39 of its members who killed themselves. They committed mass suicide under the mistaken belief that in death they would be able to board a space craft that was following the comet Hale Bopp. They were mistaken. The mainstream media picked up the story which made international headlines.

Members of a certain generation might have a problem telling you their age but have no problem telling you the name of one of the most famous comets, Halley. Halley could be the only comet known to be seen from Earth. This is a short period comet visible to the naked eye every 75 years or so. The last time it came by was in 1986. It is expected to be seen again in 2061.

Back to the composition of a comet. They are more complex than an asteroid or at the least we know more about asteroids than we do about comets. Comets have three parts, the nucleus, the coma and the tail. Again the coma could be considered its atmosphere, made up of gas and dust. The nucleus remains much of a mystery but is thought to contain ice, gas and rock. Its surface could be all rock. Its tail is formed from energy from the sun and is believed to extend for millions of miles. It is also believed that both comets and asteroids were formed from the material that was left over from the birth of the universe. But asteroids are thought to have formed in the inner solar system in the areas of the Asteroid Belt between Mars and Jupiter. Comets are thought to come from the outer solar system near or past the Kuiper Belt which is beyond the orbits of Neptune and Pluto.

After stating some of the differences between the two, if Earth had to be struck by one or the other would it not be more dignified to go by comet than by asteroid? Is not an asteroid just a rock speeding through the darkness of space? Is not a comet something mystical, illuminated by the rays of the sun that cause it to almost come to life? It can be a long distance traveler whose journey sparks the imagination as to what it has seen in the outer reaches beyond the solar system. And who can deny the wonders of its tail; made from the least substantive material in the known universe, stretching millions miles, so long it would wrap around Earth hundreds of times. A comet is so rare one can only hope to live long enough to see its periodic return, punctuated by a 75-year duration that equals the average life span of many humans. Does not the comet more so than the asteroid hold potential answers as to our human origins with the abundant water and organic material contained in its mysterious body? A comet certainly does not contain all of these answers but it can point us in the right direction, arguably more so than the asteroid.

Was it Money Well Spent?

Space exploration is certainly not cheap, coming at a great financial cost only matched by the dedication and efforts of those who devote their entire lives to advancing the science. If it is true that money spent on education is never wasted than consider the following example. A staggering 75 billion dollars has been spent so far on the International Space Station and that figure promises to grow. Sometimes we can look up into the night sky and see it without realizing it. It is moving faster than any supersonic jet as it zooms past the stars that make up its background. Construction on the station began in 1998 and it has been continually inhabited by scientists since 2000.

The fact that it is an international effort has great value when realized the worth of nations working together for the common good of the common whole. Simply put when we work together we are less likely to wage war on each other. Does this alone also not make the station well worth the fact that it is one of the most expensive science projects undertaken by humanity? Consider what might have been if penicillin, the automobile or the computer were never invented. They have all changed our worlds for the better and all were the result of scientific endeavors. The International Space Station holds the same and possibly even greater potential. Is that alone not worth the price?

Scientists study how to live in space. They research how to better travel in space. They use the station's weightlessness proximity to better study how journeys to Mars and points beyond might be possible. Up there the scientists gain knowledge about how the human body responds to and breaks down from a weightless environment, which are opportunities not afforded in the gravity laden environment down here. Scientific progress has already been made. Researchers have uncovered ways to stave off bone loss in a weightless environment and that work unveils possibilities on how to regenerate bones that have lost mass and density.

Look beyond the International Space Station's impressive capabilities and take pride in what has been accomplished. It orbits some 250 miles above us, traveling at more than 17,000 miles an hour. It is the highest and fastest home and laboratory we have ever known: the fastest cosmic condominium on record, able to travel to the Moon and then to the Earth in one day. It took conventional craft about three days to make the same journey. It's also the first permanently occupied satellite to orbit Earth.

It is called a cosmic condo because it has more living space than a large house with six bedrooms.

This craft contains so much wiring that those wires would stretch from Newark to New York City, a distance of about eight miles. It weighs almost one million pounds which is more than a large car lot with 300 automobiles on it.

Building it was no easy feat, taking a total of 136 flights to complete, only made possible by the invaluable effort of the international community. And what about its bay window with its 360-degree view, not to mention its gymnasium and ample sleeping space. Its internal pressure is similar to that of a jumbo jet. The International Space Station is truly a marvel, orbiting out there against the blue back drop of Earth, one whose true worth might still not be realized.

Curiosity and other Curiosities!

Computers have advanced our missions to and knowledge of space by leaps and bounds, one of the best examples would be probes sent to other planets. The probe Curiosity is part of a 2.5-billion-dollar mission to Mars, it's the robotic craft that landed on the planet in 2012. The rover was sent there to learn more about Mars' geology and environment and also try to determine if the planet was ever home to anything we could classify as life. It is also helping in the preparations, if the time ever comes when humans will travel to Mars.

It is about the size and weight of a car like a Mini Cooper. It required some amazing engineering to land on Mars in one piece considering the low gravity of that planet and the real chance that it would break apart on impact. So Curiosity was lowered to the Martian surface with something called a space crane that included a parachute. This amazing craft is fitted with a laser on its head that can shoot and then vaporize material. The debris from the vaporized material is then studied to determine if the landscape is toxic. The laser has a distance of 23 feet with surprising accuracy. Curiosity also has some 17 cameras that can take really small pictures, some smaller than the width of a human hair. The craft can also bore into the surface of Mars, father than other probes with its two inch drills attached to its arms.

If slow and steady wins the race than Curiosity is the clear victor. Its average speed is far less than a mile an hour down to about 0.00073 mph. But consider the fact that Curiosity has been on the planet for more than two years and traveled about 12 miles inspecting craters, rocks and the overall surface. It also has a built in nuclear power plant that is capable of generating electricity for at least 14 years. Scientists say we should all be proud of its slow, steady and dependable production.

A sixth grade student from Kansas got to name the Rover. That after the student won a contest designed to come up with a name for the craft. It traveled to Mars at a speed of about 13,000 miles an hour which took about eight months but its descent into the Mars' atmosphere was described as seven minutes of terror again because of Mar's thin atmosphere and the possibility that it would crash land.

Curiosity is not alone. There are two other rovers there, Spirit and Opportunity. Spirit is no longer functional but Opportunity, apparently appropriately named after making the most of this opportunity is still on the job after punching the clock in 2004. Curiosity could be the older sibling of the three as it is some ten times bigger than the others.

Besides collecting data and conducting experiments, Curiosity is trying to communicate to anyone or anything out there. It does this with every revolution of its tires which have been stamped with the image JPL, short for NASA'S Jet Propulsion Laboratory. It is a coded message of sorts left behind in the Martian soil that hopefully will at some point illicit a response.

By all accounts Curiosity has been a success especially when considered at last check out of the 39 missions to Mars 24 of them have ended in failure. Part of Curiosity's success is hidden in its design which allows it to roll over rocks as big as 30 inches. It landed on the planet some 14 minutes before anyone knew it, that's the length of time it takes radio waves sent from Mars to reach Earth. Radio waves blast through space at near light speed, thought to be 186,000 miles per second. It was a long distance message, blistering through some 34 million miles of space to transmit the data, which was good news by way of special delivery, so special that the sitting president instructed immediate notification the moment the message was received.

How did They do That?

How do you slow down a vehicle the size of a car, traveling at 13,000 miles an hour to a complete stop within seven minutes? Use a Sky Crane. You might now have a better idea why those seven minutes were called "seven minutes of terror" by NASA officials when Curiosity landed on the Martian surface, quite the accomplishment considering about more than half of all probes sent their crashed. It is also why Mars is called the grave yard for space craft from Earth. The Sky Crane was the apparatus NASA used to land Curiosity in one piece. The atmosphere on Mars is too thin to use air bags, which was the method used to land the two previous smaller probes, Spirit and Opportunity. In fact, Mar's atmosphere is about 100 times thinner than Earth's which means atmospheric pressure there is less than one percent of what it is here. Remember the Sky Crane was strapped to Curiosity which weighs about a ton. Earlier probes, Spirit and Opportunity weighed just a few hundred pounds.

Here's what happened. The landing process was conducted in stages and as you can imagine took years of planning. The Sky Crane and Opportunity were contained in the re-entry craft that took the initial brunt of the force of slamming into the thin Martian atmosphere. It encountered temperatures of about 1,600 degrees, hot enough to make the craft glow white. In short it used a series of reverse rocket boosts and a huge parachute to slow down from supersonic speeds to around a 1,000 miles an hour and then to about 200 miles an hour.

First the parachute was deployed, then the heat shield was detached and the rocket's reverse booster is fired up. Now less than one hundred feet above the Martian surface the sky crane lowers the rover on cables, and around 50 feet the crane severs the cables and flies away. The final stage of touch down is complete when the rover lands on the surface, which meant it went from 13,000 miles an hour to a complete stand still in a matter of minutes.

The multistage procedure contained a blind spot where the craft was still too high to use its radar. Radar was crucial in determining not only a safe landing spot but one that could be tracked and where communication would be possible. If it did not land near its original target which was a crater, who knows what kind of problems might have occurred. None did, because after the passing of the thirty second blind spot in which no radar could be used, there was enough time to make the necessary adjustments.

The system was tested and re-tested on Earth but there's nothing like the real thing. Years of effort and billions of dollars could have all been for naught had the landing failed. Failed missions can have bad repercussions that can last for years. When the Viking project failed to find life on Mars it was 20 years before NASA sent another craft to that planet. Opportunity's engineers had just one chance and NASA made the most of it. When Opportunity touched down, cords and lines were cut and the rover began its work. There's a good chance that was the last we will see of the Sky Crane but not because it did not perform well, NASA is getting out of the Sky Crane business at least for a while but in the meantime it's working on computer technology for the future and is focused on something called the quantum computer.

Quantum!

The term quantum can be defined as a reference to quantity or size and in physics it is used to measure the smallest amount of material or matter. The quantum computer is nothing like a conventional computer, so different that you cannot build one by traditional means with transistors and diodes. In order to build a quantum computer, you need new technology. This technology is so new scientists are not quite sure how to work with all of it just yet but they are close enough to already reap some of the rewards. A traditional computer could take more than a billion years to work out an extremely complex problem. It might take a quantum computer about 20 minutes to solve the same problem.

In a quantum state, particles can achieve something called superposition. It might sound like science fiction, like so much of cutting edge science but it is legitimate science. Superposition allows particles to exist in several states at the same time similar to two waves that might overlap. One example might include particles that exist in gas and in water at the same time. In this quantum state particles can enter something called Profound Entanglement, similar to Superposition, in which particles remain connected to each other no matter how far apart they are. An example might be a particle in New York and another particle in California might still be connected to each other.

The quantum computer also raises the philosophical question of artificial intelligence. That question has been pondered for years and still there is no answer. Can a computer actually achieve artificial intelligence? Those against the idea argue that a computer could never do some of the things that the human brain can. Proponents argue that the human brain is already similar in that every function and computation the brain makes can be compared to a computer. To the common layman the question seems to hinge on whether a computer can become self-aware, can it actually gain consciousness and make decisions not only based on recorded information in the form a data but also on other considerations such as morality and compassion.

Quantum is most commonly defined as a study dealing with the smallest forms of light and energy. In a strange contradiction we could learn the most about the big universe by studying its smallest parts. This is a tremendous aid to scientists who for years have looked closely at life on a sub-atomic level. And for years this question has remained: why do large objects such as birds, planets and everything in between behave in a predictable manner while really small things on the subatomic scale do not.

If and when quantum computers are better understood they hold tremendous promise for a huge machine called the Large Hadron Collider, referred to as the LHC. While many other computers are used to study the universe, the effect of gravity on humans, space travel and origins of light and heat, the LHC apparently has only one objective: to come up with a factual explanation for the origin of the universe and it's doing so by studying some of the smallest elements known to science.

Scientist with the LHC project would like to disprove the Big Bang theory of how the universe began. They are using one theory to try and debunk another theory in another effort to come up with a fact. The theory used for the debunking is called Rainbow Gravity. Through Rainbow Gravity they intend to prove that the universe was not created in an instant by the Big Bang but instead by a parallel universe.

First let's take a look at this impressive machine called the LHC. It contains four large laboratories inserted around a ring shaped tunnel in Europe. It's right out of the latest James Bond movie. This underground futuristic laboratory is so large that it travels under a number of neighboring countries. Its labs are housed in a ring shaped tunnel that spans some 16 miles and is more than 550 feet below ground. The machine is not only a lesson in speed, but also in precision. When it was being built, its two ends, again separated by some 16 miles of tunnel lined up less than one centimeter apart when the construction was completed.

The LHC accelerates hydrogen protons in opposite directions so fast that they travel at nearly the speed of light. Remember light moves along at 186,000 miles per second. The environment in which all this takes place is colder than the deepest regions of space. Those areas are around 450 degrees below zero. It is said that these high speed collisions generate temperatures that are 100,000 times hotter than the sun. The sun's surface temperature is around 13,000 degrees while its core checks in at about 27 million degrees. Either way the LHC can generate tremendous heat. The intense heat is needed in an effort to recreate the temperature just after the Big Bang, a figure thought to be trillions of degrees, maybe too hot for even light to shine.

At top speed the LHC generates about one billion collisions every second. With the help of more than a thousand computers they reduce that figure of a billion down to about 100 and it is these 100 collisions that are the most promising. The machines that study these super small, super-fast moving particles are so sensitive that their observations could be compared to recording the energy released by a flying mosquito.

The LHC also has lots of filaments which is a thin wire used to conduct electricity. These filaments are about ten times thinner than a human hair and there are so many of them they if they were lined up end to end they would stretch out 588 million miles. That's about 150 trips to the moon and back. Scientists from all of the world except Antarctica come to learn from and work on the LHC. Its magnet system has more iron than the Eiffel Tower and was completed in 2008 at a cost of about 6.2 billion dollars. We are also told all of these super high speed collisions and intense heat represent no threat to the human population above the ground, later officials will explain why they believe the LHC poses no such threat. That belief was put in writing, possibly to ease fears of catastrophe above ground in one or more countries.

Back to this rainbow theory. Scientist say if the LHC can detect the presence of small black holes it could prove the existence of a parallel universe which would disprove the Big Bang Theory. A black hole is a region of space where gravity is so strong that not even light or any form of matter can escape. Black holes are created by the collapse of a very large star and are millions of times larger than our sun. Under the Rainbow Theory the universe would stretch back into time with no beginning that we can detect and consequently no Big Bang. They say the end result would reveal the existence of a parallel universe. Scientists say they can calculate the speeds required to detect these small black holes in Rainbow's Gravity and if they can do it, then they say their theory would be correct. After a lengthy shut down for upgrades research continued with the LHC now almost twice as fast as the original.

If parallel universes do exist it is said that you could compare them to two pieces of paper. Gravity from one universe would seep into the other. The creation and identification of a black hole, in this case a miniature one made in the LHC, is expected to display this process. Only problem is if it works than what? Are we dealing with at least two universes that pose twice as many questions than answers? We have just begun to barely understand the first universe, never mind a second, third or who knows how many more universes could be lying in wait. The human mind understands concepts in two and three dimensions. One universe has many dimensions said to be well beyond human comprehension. If we cannot fully understand one universe what makes us think we could understand many. Therein lies one of the beauties of science, the notion about discovering as much as possible, even if we can't completely comprehend it.

Those for and against the Rainbow Theory would probably agree with this proposition: anything of substance occupying the universe and was the product of a collision or explosion. This would include the Sun, the Earth and the nearly countless number of solar systems and galaxies that comprise the universe. Is the LHC on the verge of something really big, only made possible by studying something really small? Will the LHC turn theory into fact? If and when we fully understand quantum computers will they be responsible for such a discovery?

Fact is defined as a thing this is indisputably true. Science is defined as the intellectual and practical activity encompassing the systematic study of the structure and behavior of the physical and natural world though observation and experiment. If LHC does come up with this little black hole would humanity then have a tangible black hole to test and retest in an objective manner free from bias, theory or theoretical interpretation. At this point that is something that the Big Bang Theory cannot offer.

According to the experts, including Albert Einstein and the Large Hadron Collider Safety Assessment Group the LHC cannot produce a black hole because a black hole, considering all of the galaxies, stars and planets it has consumed is just too big and heavy to reproduce in a laboratory. So can the LHC create a smaller black hole? Apparently the answer is still no. Reason being the LHC mainly deals with really small sub atomic particles which have a tendency to disintegrate immediately. Black holes in space need time to grow during which they attract material, such as galaxies in what has been called an accretion process and any hypothetical black holes created in the lab would immediate fracture or decay into nothing in part because they would be so small. This was part of the explanation put into writing in order to ease fears of an above ground catastrophe.

What's the Relationship?

Is there a connection between the Large Hadron Collider and something call the God Particle? The short answer is yes. But first things first. We're told it would be bad form to call this piece of sub atomic matter the God Particle. Instead it would be more proper to call it the Higgs boson, named after the person who discovered the idea, Nobel Prize winner Peter Higgs. But Mr. Higgs doesn't believe in God, in fact he's wholeheartedly a devout atheist. So it's probably a bad idea to call this theory the God Particle when it was put forth by a person who doesn't believe in God. In fact, Mr. Higgs has bluntly asked people to stop using the term the God Particle, saying it's not funny and is misleading. The term Boson is a reference to a physicist who had the last name Bose.

With that understanding, the Higgs boson is thought to be the sub atomic particle that gives mass to all matter but the problem is, it's never been discovered. Many of the top minds believe that there must be a Higgs boson element that pervades the entire universe and interacts with all other known matter. It is a crucial part of a standard belief regarding particle physics, which is the study of the behavior of some of the smallest particles known to science. Again, no one has ever found it. But the LHC is designed to discover it by attempting to recreate conditions immediately after the creation of the universe. We are told that despite its tremendous power, the LHC won't blow up part of the planet and that promise has been put in writing as we recently learned.

The LHC is a reflection of the human character. Humans differ from most other forms of life in that we are curious, we possess an intellectual curiosity that has taken us from the cave to the moon but yet we still don't know where we came from but the LHC might have the answer. It's thought the Higgs boson is much larger than most other sub atomic particles, maybe 200 times the size of a proton. The larger the collider, the larger the collision and the better the chance of producing and identifying a miniature black hole, which is one of its major objectives and why the LHC is so gargantuan. When these particles collide they leave behind debris, including possible debris from a black hole. Another problem facing scientists with the LHC is deciding when they have the required proof. Do they announce results after one successful experiment or will that announcement come after several experiments? We are told they are happy to solve the problem once it appears.

Seems the LHC can produce in a laboratory what nature currently cannot manufacture in the universe and that is the hottest known temperature. In the process of searching for the elusive Higgs boson particle, gold particles are smashed together at speeds of near 186 thousand miles per second, equal to the speed of light, so fast that they reach 7.2 trillion degrees Fahrenheit for just a split second. It is believed that was close to the temperature of the universe right after the Big Bang. That is hotter than anything known, including a super nova and about 250 thousand times hotter than the hottest part of the sun. We are told things can get even hotter, but why try and explain what we can't understand it. Suffice it to say at those temperatures strange things happen and the natural laws as we know them break down.

Finally, would the LHC have been created had there been no Higgs boson mystery? Considering the nature of serendipity and the pace at which science advances, the answer might be yes. Serendipity is commonly understood to be the occurrence and development of events by chance in a happy or beneficial way. Or put another way good things were discovered by mistake. Ponder some of the positive results of serendipity, penicillin, the microwave, ice cream cones and champagne and the pacemaker have all improved humanity. Will the LHC soon be added to the serendipity list by discovering something good that was unrelated to the Higgs boson?

So close but Still Too Far!

Scientist were excited after they recently discovered a new planet, the elation comes from the fact that the planet promises to more like Earth than any other they've seen so far. It's too far away to be in our galaxy so it's too far to visit at some 39 light years away. So the planet is close by galactic standards but an impossible distance to traverse, putting the distance of this new planet at 234 trillion miles. But just because we can't go there does not mean we cannot learn about it. Observations from super powerful telescopes promise a wealth of new information. These telescopes are so strong and the new planet is not too far that astronomers hope to see if it has an atmosphere. So the excitement is genuine and with good reason, every knew discovery just enhances our knowledge and this one is no exception. Space telescopes seem to provide what space travel cannot and that is a wealth of information about anything farther than the moon.

The new planet is called GJ1132b. Here's what we know so far. It's too hot to contain water or support life as we know it. Its surface temperature is about 400 degrees Fahrenheit. That's about as much heat as the average oven in your house can crank out, which eliminates the possibility of finding any life as we know it on the surface. Its home is in the constellation Vela which can be seen at the right times in the southern sky. It is rocky and about the size of Earth and it orbits a star called a small red dwarf. Red dwarfs are relatively cool stars and are the most common stars in the Milky Way but they're not real bright and are hard to see and this one is about five times smaller than the sun.

GJ1132b is about 1.4 million miles from its red dwarf. Put another way, it is much closer to its star than Mercury is to our Sun. GJ1132b is so close to the red dwarf that it is tidally locked in its orbit, which means it does not spin on its axis like Earth and other planets. Earth spins every 24 hours which is calculated by the passing of one day. Because GJ1132b orbits its red dwarf so quickly its year could be compressed into a day and a half. So the planet's year doesn't even last two days. Still scientists say this could be the most important planet found outside of the solar system. They have already scheduled time on the Hubble Space telescope to have a better look at it. The planet could also be the target for future missions not involving humans but instead with telescopes. This includes the James Webb Space Telescope which is scheduled for launch in 2018.

Some three months earlier another new planet was discovered. This one was not nearly as close and it is much bigger than Earth. But this planet actually has a name instead of some sort of coded combination of letters and numbers. It is called Kepler-452b. OK it does contain three numbers and one letter but at least it has a name. Kepler-452b is about 1,400 light years from Earth, too far to try and see if it has an atmosphere but close enough to see it's about 60 percent larger than Earth. Which means its gravity might be about twice as strong as Earth's, probably too strong for humans to survive without significant adaptation.

Kepler-452b might be about the same distance from its star as Earth is from the Sun, estimated at 93 million miles. Its atmosphere is about twice as thick as Earth's and there is a good chance that it has a rocky surface with possible volcanoes. Its year is some 20 days longer than ours but because it's been around longer there is the possibility it has had enough time to produce life, if basic conditions were met. Which each new discovery of another planet our understanding of the universe grows but so do the number of questions. On average one new planet is discovered every day. Which leaves one with the inescapable fact: the more we know the more we must learn.

The Genuine Articles!

―――――――

Most of us have heard about shooting stars and might think about comets but the real ones are just that, shooting stars called hypervelocity Stars. Their names give away much of what they do, hyper means unusual energy and velocity refers to speed, so to say they have unusual speed is an understatement. They're huge, fast and super-hot, zipping through galaxy at millions of miles an hour. Scientists believe that at one time they were part of a binary star system. But when something happened to the companion star, like getting consumed by a black hole, the other star was ejected at a very high speed into the outer reaches of the universe. Did we mention that they are about four times as large as our sun?

These stars are rare, out of the billions of kinds of stars in the universe only about 20 of these special ones have been found. So 20 out or more than a billion is rare and these hypervelocity stars can tell us a lot about the inner part of the galaxy and the universe. It's also hoped that hypervelocity stars can shed some light on dark matter, which is the mysterious substance that surrounds some of the brightest areas of the universe. The first one discovered was aptly named HVS1, for hypervelocity star the first. They are ejected out of the galaxy about once every 100,000 years, which might explain why they're so rare. Whenever a new star or planet is discovered it promises to tell us something new about the universe and that's the case here.

There is something even faster that a shooting star, it's called a runaway star. One such runaway star is racing through our galaxy covering thousands of miles per second or more precisely 2.7 million miles every hour. It currently holds the record for the fastest star in the Milky Way. Our sun, traveling around 483,000 miles an hour is not even a close second. This one is called US 708 and it is going so fast that its speed will overcome the gravitational pull of the galaxy and this run away star will run right through the galaxy and into the great abyss that lies beyond. How did US 708 get up to speed? Scientists don't really know but think it got up to speed by getting ejected from a black hole. Periodically a black hole might provide answers instead of raising questions. When you consider the tremendous gravity of a black hole, so strong that even light cannot escape, it seems understandable that such a force would be capable of generating such a speed. Researchers were able to track its speed by a series of observations that took 70 years.

It is believed that US708 also has different origins that many other stars. The term "different origins" cannot be pinned down much further other than to speculate that the star's origins were chaotic and strange. They also speculate that it was probably a part of a binary system with a companion star. The two stars could have been orbiting each other high speeds and close proximity. Somehow the path of the companion star might have changed and brought it too close to the black hole at the center of the universe. So it was already whipping around its companion star and unbelievable speeds when it was caught at the right time and at the right angle by the black hole's super strong gravity when the companion star exploded and went super nova. A super nova increases a star's brightness because of a catastrophic explosion that ejects most of its mass. So maybe these two forces combined to get US708 up to speed, traveling again at 2.7 million miles an hour or at least so the theory goes. Where's there is one there's usually more and that's the case with runaway stars. Another runaway star was recently discovered near something called the 30 Doradus Nebula. Nebulas are huge clouds of gas and dust in the outer reaches of the galaxy and are often called star nurseries.

Not all of these runaway stars are moving at more than a million miles an hour. In comparison some of the stars are plodding along at 250,000 miles an hour. A run away star might first appear in the day time sky as a small spec that grows rapidly on the horizon and soon appears to be the size of the sun. The fact that there are two stars in the sky might be enough to cause considerable panic. Under this scenario the day time sky might be much brighter and increase glare to uncomfortable levels, temperatures might go up to dangerous levels and the rates of skin cancer could increase to deadly levels. At 2.7 million miles an hour the runaway star would rapidly approach Earth and its true size would soon be revealed, increasing panic, heat, and the blinding glare from its cancer causing radiation. Depending on its trajectory the runaway star might shoot off into the day sky just as rapidly as it appeared, making one big arc across the horizon like something out of a cartoon. If it were on a collision course things might not be so funny. The aforementioned increased heat, panic and the blinding cancer causing glare would grow until the runaway star was slow close that it covered the entire day time sky. At close proximity the oceans might evaporate and combustibles on the surface including humans would be incinerated. If this scenario happened at night, night might turn to day.

There's a planet from hell called Gliese 581C and it would kill you. The planet orbits a red dwarf star in close proximity which means Gliese 581C is tidally locked in place. That also means the planet does not spin on its axis like other planets and stars. Gliese 581C always shows its same face to its star which can create some strange conditions. On one side of the planet the heat would melt your face and the other side would freeze was left. So hot and cold all at the same time, it is literally a planet of burning ice and it is moving at more than a million miles an hour. Fact can be stranger than fiction.

So why doesn't this burning ball of ice just melt? Scientists think that's because Gliese 581C has water underground and is drawn toward its core by intense gravity. That gravity is so strong that it pulls the water molecules together with such force that they cannot evaporate. Gliese 581C is about 20 light years away. If its name rhymes with grease, then you know you have the correct pronunciation. Its year is a little more than a month, taking some 37 days to complete its orbit. Gliese 581C's mass is about four times that of Earth and weighs about two or three times as much. That means if you weighed 120 pounds here you'd weigh 360 pounds there. It has no moon and no life and would take a long time to get there. Imagine if humans could travel at one tenth the speed of light it would still take about more than 200 years to reach Gliese 581C.

Imagine a planet that is a girl's best friend, well it might be a diamond. That was part of a popular advertisement in the previous century. It was used to sell diamonds and went something like "A diamond is a girl's best friend." Well, if the universe is home to everything under the sun, then it's also home to planet that is a girl's best friend, that's right a planet made from a diamond. Instead of calling this newly discovered planet some combination of letters and numbers, why don't they just call it Diamond. Wonder how much it would cost! Anyway, it was once a part of a binary system. But at some point its companion started to consume its partner planet and in the process increased its value, at least in the eyes of materialistic humans. Scientists say that initially the two planets were together when one began to consume the other. The core of the partially consumed planet was exposed and when carbon, which was in the core is exposed to enough intense heat and pressure that carbon is turned into a diamond, which is how a planet can be a girl's best friend.

While diamonds on Earth are rare, they might be more plentiful in the universe. There is another one these diamond planets called Cancri e. It is about twice the size of Earth but weighs at least eight times more. A third of Cancri e's mass could be made of pure diamonds. That's nearly equal to Earth's total mass, which that begs the question if diamonds are so plentiful are they still valuable? Much of Earth's surface is covered in dirt. Some might say dirt is so plentiful that it's worthless. So could diamonds on one of these planets be so plentiful that they too are worthless? And if you only had a year to live you would not want to spend it on Cancri e. That's because its year is less than a day, taking just about 18 hours to complete its orbit. Another reason you would not want to spend much time there is its heat. Temperatures on Cancri e reach a scorching 4,000 degrees Fahrenheit, so hot that 18 hours might seem like a year.

While water in one of the heaviest substances on Earth this is not so in space. In fact, there is a huge reservoir just floating out there. The size of this reservoir is estimated to contain more than a trillion times the amount of water in all of the oceans, lakes and rivers on Earth. It is called the largest reservoir in the universe. But it is not like a reservoir that one might think of on Earth. Water turns to gas or vapor in space because there is not enough pressure for water to retain its liquid form. This reservoir is vapor, a giant cloud of gas that is several hundred light years in diameter and about some 12 billion light years from Earth. The giant reservoir is in the heart of a quasar which is fueled by a black hole. From our understanding, a quasar is short for quasi-stellar radio source and is thought to be a remote object emitting large amounts of energy which might look like a star. Other scientists say they might be series of galaxies getting consumed by a black hole and in the process they shine brighter than anything in the universe. Yet other scientists say they might represent a stage in the evolution of galaxies. So we are not quite sure what they are. But water vapor might reveal the nature of these quasars which indicates that the vapor existed early in the universe. This water vapor could have existed for more than 12 billion years in the form of this huge gas cloud, so the potential to learn from it is as enormous as its size. The vapor was discovered by two research teams on two mountain tops half way around the world from each other and it was their combined measurements that confirmed water played an early role in the evolution of the universe.

Who Knew!

———————

There are many unsung heroes whose sacrifices and accomplishments don't get the attention or credit they deserve. What comes to mind are the many early high altitude test pilots and early astronauts who paved the way to the moon and beyond. As a boy I remember surfing through the few channels available at that time and thought I might be watching a man die. The actual event happened years before the broadcast but the man who was central to the story was a high altitude military sky diver. The man had just landed and lay there on the white sand of New Mexico, incapacitated. He seemed to be choking to death and his hands appeared to be around his neck as he gasped for air. Also one of the man's hands seemed to have swollen to grotesque proportions. If I remember correctly there was another man there who stood above him but he did nothing, possibly frozen with fear as to what might soon unfold. The picture was interrupted but moments later the image returned. The same man was still there on his back in the sand but now he was laughing with others who just arrived and they exchanged congratulations. It was hard to imagine that one moment he appeared to be near death and the next all was seemingly forgotten and replaced with carefree joy and laughter among friends.

That was Colonel Joe Kittinger and this is part of his amazing story. The year was 1960 and the military was conducting high altitude parachute jumps. At the time the exercises were said to be for the benefit of military fighter pilots who might be shot down at high altitude. But others speculated it was an effort to see if a person could actually fall from space. At the time NASA had not yet been created but there was great military interest in trying to somehow get into space. Very little was known about the area below space but still well above the Earth. This area covers a distance of 62 miles above Earth which is the accepted definition of where space actually begins. This 62-mile area was nothing more than a sliver when compared to the vastness of the solar system soon to be discovered where distances are measured in light years. But even at an altitude of 63,000 feet human blood begins to boil. Another 30,000 feet up and the temperature plummets to 90 degrees below zero. At such a height there is not enough pressure to hold liquids together so human eyes could also boil away. The American military wanted to learn as much as possible about these altitudes and decided to send sky divers to these heights as test subjects.

Joe Kittinger willingly signed up for the mission. Like many test pilots he was a dare devil. As a young man he raced speed boats. So he had no reservations or at least did not show them, when he jumped into the basket beneath a special balloon and soared to a height of 100,00 feet and used his parachute to float back down to Earth. The big problems associated with the low pressure of the stratosphere soon became evident. Joe Kittinger, who was a captain at the time, experienced some sort of rupture in the pressurized suit on his right hand. Not knowing what to expect from the unknown, Captain Kittinger proceed with the mission. At this point he was 50,000 feet up and with that right hand exposed to the vacuum in the upper atmosphere it began to swell. But he continued with the belief that he could survive the increasing pain in his right hand. That hand would eventually enlarge to grotesque proportions. There was no explanation as to why he seemed to almost choke to death but he survived and again, moments after the close call seem to laugh the whole thing off with military personnel.

Who knows what must have been going through his mind considering one year earlier he was almost killed in similar fashion. In November of 1959 Captain Kittinger's stabilizer parachute opened prematurely and the cord wrapped around his neck. Captain Kittinger passed out but the main chute opened and the stabilizer chute broke away. There was an unconfirmed the story of another high altitude military sky diver that year who survived the jump from the edge of space but drowned after he landed when he slipped out of the rescue basket and fell into the ocean.

In fact, Captain Kittinger had three high altitudes jumps before he broke the record. I guess the third time really is the charm. That third time he floated to a height of 102,800 feet or about 20 miles above Earth. Captain Kittinger said he had no sensation of falling and would have to turn around to look at the blackness of space to reorient himself. In this paraphrase he said he could not visualize the speed, no depth perception, like driving down the road with your eyes closed, there's no 600-mile an hour wind blowing on your face, he could only hear himself breathing inside the helmet. His comments were said to prove a theory put forth by Albert Einstein a few years earlier regarding human sensation in an environment with little or no gravity.

It was no coincidence decades later in 2012 that Joe Kittinger, now a Colonel was selected to head mission control for dare devil Felix Baumgartner. That was when Mr. Baumgartner broke Colonel Kittinger's sky diving record. Colonel Kittinger was now 84 years old. Few people knew that he also was the first man to make a solo crossing of the Atlantic Ocean in a gas balloon. Years earlier when Joe Kittinger was fighter pilot in Vietnam his jet was shot down and he spent about a year as a prisoner of war in the camp called the Hanoi Hilton. His cell mate next door was John McCain. Mr. McCain would go on to an esteemed political career that included lengthy terms in congress and a presidential campaign. Colonel Joe Kittinger's accomplishments are too long to list, but what a life, from a teenager who raced speed boats to arguably the first astronaut. The vast majority of his work advanced our knowledge of aeronautics and helped to lay the ground work for space travel for years to come.

Walking at 18,000 Miles Per Hour!

The story of Ed White is much shorter, possibly because Ed White had a much shorter life. He was the man with the common name but with the uncommon accomplishments. I first read about him after his spacewalk. The big bold headline read Ed White completes the First American Space Walk. I thought wow! What's a spacewalk? The article did not give a straight forward answer as to what a spacewalk was. If my recollection is correct it read something like, he floated 100 miles about Earth and traveled from Hawaii to Bermuda in 20 minutes. That's a distance of about 6,000 miles. That meant in space he was walking around 18,000 miles an hour. On Earth the average walking speed is 3 miles an hour. I just had to see a picture of this but that would have to wait. I finally saw the picture that amazed so many folks, there he was in all that splendor, walking in space at speeds too fast to imagine.

Ed White was floating in space. His posture resembled a person sitting in a recliner. His legs were bent as if he lounged in a Lazy Boy chair and there was a lose strap that just floated in space, just like Mr. White. He had a small hand held rocket booster that help him stay oriented and properly positioned. I thought he's not walking, he's lounging, 100 miles up, blasting through space at some 18,000 miles an hour. I think he was quoted as saying something like "I feel red, white and blue all over." The sun shone like I have never seen before, reflected off the pitch black face shield of his helmet. I learned years later that Ed White so loved his spacewalk that he would stay out there past his allotted time and had to actually be ordered back inside the Apollo space craft.

Edward Higgins White II was born in Houston, TX in 1930. He was a boy scout who later was accepted at West Point. His was gifted with both exceptional academic and athletic skills. He nearly became an Olympian, missing the US Olympic team by just $1/10^{th}$ of second. He ran the 400 meter hurdles which is one of the most demanding events. Years later he studied aeronautical engineering and received a Master of Science degree before becoming an elite test pilot in the US Air Force.

Lt. Col. Ed White has at least a dozen memorials in his name and received as many awards. He has at least 8 schools named after him in Texas, Illinois, Iowa, Florida and Alabama. He was also featured in at least two big screen movies that I can think of, Apollo 13 and the HBO series, From the Earth to the Moon. Mr. White was not only the consummate astronaut but he was also a devoted husband and father of two children.

Lt. Col. Ed White was killed during a training accident in 1967. The official cause of death cited asphyxiation and smoke inhalation. But after reading part of that accident report it sounded as if he was burned alive. There was a fire inside the capsule and the exit door could not be opened. When Lt. Col. Ed White's body was recovered his arms were still reaching upward from when he tried to open that latch. His wife later remarried but committed suicide after health complications. Eight months after his death the US Postal Service issued a stamp in honor of that spacewalk, again which was the first ever by an American. It was also the first time that the postal service used such a design which took up two stamps but his name did not appear on the stamps. Ed White was also issued a congressional medal of honor for his aeronautic accomplishments after his death.

Like so many before and after him Ed White's inspiration extended well beyond his achievements in the US Air Force and with NASA. The life he led was exemplary on every level. Considering what he was able to accomplish in a life cut short only begs the question what could he have done if he had more time.

The Renaissance Men!

I never heard of John Stapp until I was an adult but certainly benefited from his accomplishments just like everyone else who travels by land or air. If the dictionary defines a renaissance man as a person with many talents or areas of knowledge, then it should have a picture of John Stapp next to its definition. He was born in 1910. He was a medical doctor and flight surgeon, a biophysics expert, a jet pilot in the military, a researcher and a pioneer in studying the effects of acceleration and deceleration. He had degrees from all over, including a bachelor's degree from Baylor, an MA from Baylor, a PhD from the University of Texas, an MD from the University of Minnesota and an honorary Doctor of Science, also from Baylor. If all that were not enough, John Stapp was also the fastest man on Earth.

Some of his amazing work included testing oxygen systems in jets with no pressurized cabins flown at altitudes of 40,000 feet. The tests were designed to learn more about decompression sickness, commonly known as the bends. Decompression sickness is a condition when dissolved gasses are released from the body into joints, muscles, organs, and just about anywhere else in the body. It can cause extreme pain and even death, brought about by rapid and drastic changes in pressure. At the time one of the problems with high altitude flying was pilots would get the decompression sickness. John Stapp's work solved this problem. But his work on increasing travel safety was just beginning.

In the 1940's much of his work concentrated on the effects of deceleration where early tests were conducted on rocket sleds. These contraptions would speed along the track at more than 500 miles an hour. One of the first ones ran off the track. At the time the rocket sleds were not manned, maybe because they would run off the track. But a short while later the rocket sleds would have human pilots. By 1950 some 74 human runs had been conducted with Mr. Stapp being the most frequent volunteer for these high speeds thrill rides. Many of us have seen pictures of the pilot's or passenger's faces with mouths open, faces stressed, and the skin on the faces rippled back by a series of high speed vibrations similar to a flag blowing in a stiff wind. Well, that's what John Stapp liked to do.

Despite breaking bones he was not deterred from his work, some of it concluded that a person could withstand greater deceleration forces if he or she was seated in the backward facing position. This position also required very little harness support. Mr. Stapp's work proved this beyond a shadow of a doubt. The military stood up and took note and installed new backward facing seats and retro fitted the existing ones on its aircraft. But for whatever reason the commercial airline industry ignored the findings.

He rode the rocket sled a total of 29 times and on his final ride he reached a speed of 632 miles an hour. That effort broke the land speed record. John Stapp also sustained what could have been record number of G forces proving that the human body can withstand at least 45 G's. One G is the force of Earth's gravity. At 5 G's a person experiences a force equal to five times his or her weight. Put into context you have this middle aged man, speeding along the ground at more than 600 miles an hour with the equivalent of 45 other men of the same weight all sitting on top of him. He loved it. Remember he did it almost 30 times.

As you can imagine John Stapp conducted extensive research in the sky as a jet pilot. He contributed greatly to something called wind blast experiments. The tests were designed to recreate the forces and pressures present during high speed accidents like when a canopy blows away from a jet. The tests would also determine if the pilot could stay with the aircraft without the canopy. Mr. Stapp stayed with the craft at a speed of 570 miles an hour and didn't suffer any injuries from the extreme wind blast. He also took part in one of the first high altitude skydives.

Of course it should go without saying that he was a great thinker and loved to described the world around him with language that require extensive thought. He would express general truths through astute observation, using the spoken and written word to made profound distinctions and definitions. Some of his work includes: "life is short, art long, opportunity fleeting, experience deceptive, judgment difficult." In the early 1990's he published a collection of some of his work. He was also credited with being the author of the final form of Murphy's Law, "anything that can go wrong, will go wrong." And what about Stapp's Law: "The universal aptitude for ineptitude makes any human accomplishment an incredible miracle."

The work of John Stapp and his willingness to volunteer for so many tests and experiments not only improved the safety of air and ground that no doubt saved countless lives. Those first flights into space were a success in part because of John Stapp's work. But I would not be surprised if John Stapp viewed his work as fun, comparing the rocket self to a youngster on a roller coaster. His face was so contorted by the G-forces at 500 plus miles an hour that it was unrecognizable, but the moment the force stops, John Stapp could be seen smiling.

It is comforting to know that John Stapp was not killed in a high speed accident or in a fire like many of his contemporaries. At the ripe age of 89, he almost lived to the see the year 2000, but died just one month shy, passing away peacefully at his home in New Mexico.

Society might expect great accomplishments from the aforementioned men who were all exceptional and had extensive histories in the military but what about a rock and roll star. Brian May is a man of many talents including a musician and a scientist. For years he was the lead guitarists for the rock group Queen. He co-founded the band and has been called one of the greatest guitar legends of all time. Now he's devoting more time to astronomy which is his other passion. In 2007 he attained his PhD in astrophysics and in 2015 he spent a number of days with the New Horizons team. That's the group of people who helped send the New Horizons probe by Pluto and other distant places in the solar system. He is also a co-founder of Asteroid Day, which is day designed to promote awareness of the dangers posed to Earth by asteroids and comets. The first Asteroid Day was in 2015 and this year's event promises to be more successful than the last. It is a time when the European Space Agency and other people will discuss the current technology of searching for asteroids and some of the best ways to avoid an impact.

What Could Have Been!

With the exception of 911 and major wars the most devastating moment for our nation might have been the Challenger disaster. So much was riding on what was supposed to be an almost routine shuttle flight considering how many previous flights had taken off and landed without incident. A teacher was on board that flight January 28, 1986. A civilian riding the shuttle into space gave hope and inspiration to millions of Americans who seemed to be growing bored with the space program. So many people thought if a teacher could transcend her destiny and enter space, maybe they could too.

Just seconds into the flight disaster stuck, an explosion occurred and everyone on board was killed. Arguably, a record number of Americans watched the flight and ensuing disaster, again possibly because a teacher was on board. An estimated 17 percent of Americans were witnesses. Just like 911, many people remember exactly where they were when the Challenger broke apart. On the night of the disaster the president was scheduled to give the State of the Union address but postponed it for a week and opted instead to sooth, calm and reassure our country by giving a national address on the disaster from the Oval Office. NASA had hope it's Teacher in Space Project would increase national interest in its space program but the Challenger explosion derailed those ambitions.

That teacher was born in 1948. She was selected for the program out of some 11,000 applicants and had planned to use her experience in space as a learning tool for her students. Instead those students watched her die in a fiery accident. That disaster resulted in a

32-month hiatus for NASA as it struggled to get its house in order. The investigation into the accident revealed that potential deadly flaws were identified years earlier but were either ignored or just not acted on. Apparently there was an ''organizational culture'' to push ahead regardless of the known dangers that were present. So much for generating a renewed national interest in the space program.

Possibly more gruesome than the actual accident was the recovery of bodies. Their charred craft slammed into the ocean at more than 200 miles an hour. That craft remained submersed in sea water for weeks, exposed to the corrosive effects of salt and scavenging marine animals, which made the victim's bodies unrecognizable. The teacher's body was the second to be retrieved. The remains of all of the victims were in a quasi-liquid state that in no way resembled a human being. Military divers refused to continue their retrieval work until the craft was hauled onto the deck of a ship because of the space craft's lethal form that contained too many sharp and jagged angles to safely continue with the process.

There was also on an engineer board. He was born in 1944 and was another civilian that gave hope to the masses that they too could transcend their destinies. During the recovery process, what was left of his body floated out to sea and was lost to recovery crews. His body was seen the next day floating on the surface but it sank before crews could reach it. There was an astronaut who was determined to avoid such an ending to what already was a horrific chain of events. So he decided to search for the engineer's body on his own and with his own money. He used a fishing boat to continue the search and weeks later dive crews were able to locate and retrieve the body which was at the bottom of the ocean about one half mile from the crash site. So ends this story of the Challenger disaster. A total of seven people were killed and along the way part of an American dream died with those courageous souls. If there's a lesson to be learned, other than progress demands a high price, many people are waiting to be instructed.

Was the American public, eager for space travel, discouraged by the tragedy of the Challenger disaster? Many people believe that putting civilians into space was one of the best ideas NASA ever had and that it was correct in trying to increase the nation's interest in space travel. A teacher and an engineer would have been just the right "ordinary people" to promote that national interest. The two pioneers must have known and accepted the risks and dangers associated with the mission and obviously they were not deterred. Should the American public not be deterred as well? Now that the project has ended a question has risen, if the teacher and the engineer could speak from the great beyond what would they say? Would they be strong advocates of continuing the Shuttle program? Would they want their deaths to derail the program they were so passionate about? Would they want NASA and the nation to finish what was started?

A teacher who was a finalist for the mission and who watched it explode said she never lost her love for space. Soon after the disaster that teacher was back in the classroom, comforting her students and putting the accident into some sort of context. That was the sentiment from many finalists who viewed the victims as having the short lived opportunity of fulfilling the lifelong dreams of the nation. More than a quarter century later that teacher's convictions remain unshaken.

In 1998 NASA replaced the Teacher in Space Project with the Educator Astronaut Project. In 1990 NASA canceled the program all together. Family members of the Challenger victims created the Challenger Center to make sure those deaths were not in vain. They accomplished this by carrying on the original mission statement of educating the public about the wonders of space. The center has some 48 learning facilities around the country. Among other things, these facilities allow kids to learn about space through simulated missions into the solar system.

Teachers in Space was recreated in 2005 and sponsored by the private sector. NASA seems to have turned the page on this amazing and vastly popular project. The American people have also seemed to have opened a new chapter with some of the aforementioned programs and facilities put in place after the disaster. This could be seen as proof that the nation has not given up on the idea of space exploration that includes the participation of its civilians. We can only hope these new programs attain the objectives set forth by the original Teachers in Space Project. Think of the possibilities and what it could do for our collective metal condition. Think about the conversations around water coolers. Public interest has always been there when you consider more than 11,000 people signed up to take part in the initial program. Think about all of the young adults who might have pursued careers in the sciences as a result of the Teachers in Space Program and their how accomplishments are waiting to benefit society.

When we look up at the heavens at night we might have an entirely different mindset. Those stars and that moon might not seem so distant, not with the knowledge that ordinary people have already been out there and come back to Earth to share their stories. The people telling those stories are not the typical astronauts who have trained their entire lives for these moments and sometimes recall them as dryly as a police officer might describe an arrest. No, these stories that would be literally out of this world would be told just like a friend telling a story to another friend. They would be full of colorful words, maybe an expletive or two thrown in for good measure, they certainly would inspire imagination and add a richness to our daily and sometimes boring routines.

How Much is Too Much?

The term Space Tourism might be misnomer when you consider the vast majority of tourists could not afford to travel in space. Some of the tickets for space tourism cost twice as much as a nice house. At last check some 536 people have traveled to space. Not a very good record in making space travel accessible to the masses when considered there are some 7 billion people on the planet. Even more dismal is the price many of these civilian space travelers had to pay for the experience. It was common for the first trips to cost 20 million dollars, making the privileged experience available only to the extremely wealthy. One trip was as high as 35 million dollars, paid by the seventh space tourist who purportedly said the price was a bargain. Probably not what NASA had in mind when it initiated the Teachers in Space Project. Space tourism is understood to be just that, space travel for recreational purposes. The end of the space race might have been brought about when humans landed on the moon. That result could have been a reduction in public interest which led to less emphasis on space exploration which lead to a reduction in public funding for the projects.

Russia had been one of the leaders in the space tourism competition but reduced its effort after the Challenger disaster, then made further cut backs during its economic crisis. Russian efforts were even further reduced in 2010 because of the demands of the Intentional Space Station which subtracted room for civilian opportunities. Just three years earlier space tourism was seen as one of those emerging markets that would develop affordable and accessible space travel for the common person. The potential for this market has not come to fruition. In fact, no sub orbital space tourism flight has yet to take place.

These kinds of flights would peak at an altitude of some 100 miles above Earth. That is well above the 62-mile level, commonly accepted as the beginning of space. At this altitude passengers, would experience up to six minutes of weightless. They would have a view of the stars unlike that on Earth, which would be free of glare and twinkle. They would also look at Earth in all, or at least in most of its curved glory. It has been said that viewing Earth's curvature from such a height leaves an indelible impression that can forever change the way a person thinks. Price tag for such an experience was approximately 200,000 per passenger.

In 2005 the United States Government proposed rules for space tourism. It took a model similar to the one used in American air travel. Screening procedures and emergency teams were part of the proposal but there were no health requirements. The exclusion of health requirements was a possible indication of its early intentions to make space travel accessible to the ordinary person. Half a decade later, New Mexico legislators mandated a law that included legal protection to the companies that provide these flights in the cases of accidental harm to, or death of its passengers. Gross negligence and willful misconduct are excluded. Passengers would also sign waivers that they acknowledge the inherent risks brought about by their own accords.

Space tourists would also be expected to help with scientific experiments. According to one researcher, public reaction to the proposition is favorable with a vast majority of those people wanting to travel to space. Almost as many people indicated they wanted to spend at least a couple of days in orbit. The criticism is that this is only for the super-rich. Space tourism has spawned speculation of reality shows, one show would even host a contest to win a flight which would literally be a prize out of this world. Economic projections are impressive. Space tourism could become a billion-dollar market. But it won't happen overnight. It could take 20 years. Some companies have reportedly recorded 80 million dollars in deposits. If you ever wanted a sneak peek at the future this could be it but it's going to be expensive.

All Aboard!

There is already a company billing itself as the owner of the world's first commercial space port. The name spaceport probably comes from the word airport. The Spaceport America terminal looks like something out of this world, appropriate considering it'll offer flights that literally live up to the billing. It was designed to blend in with the wide open desert surroundings of New Mexico. It encompasses 27 square miles and has a 12,000-foot runway. It was built by Virgin Galactic and costs more than 200 million dollars.

It is designed to accommodate sub orbital tourist flights and has a launch pad for unmanned rockets. Part of its advertising slogan is "at last, space travel is open to all of us." But critics would say that's not true because not all of us can afford tickets that cost 200,000 dollars. Still it boasts some 700 future astronauts from an estimated 50 nations around the world. Some of those future astronauts will be women who'll become the first females from their countries to enter space. The ages of the passengers will range from 10 to 90 years old. Another one its advertising slogans is exploring space makes life better on Earth but it fails to explain how.

All this was the brain child Richard Branson, an English businessman whose worth net worth was estimated at 5 billion dollars at last check. In 2016 Mr. Branson was 65 years old and owned more than 400 companies which are under the umbrella of the Virgin Group. At 16 he started his first business. A few years later he opened a chain of record stores called Virgin Records. And the rest they say is history.

Jeff Bezos is another businessman interested in space tourism. He founded Amazon and the space company backed by Amazon called Blue Origin LLC. Mr. Bezos could be stepping out of the shadow of Mr. Branson with recent developments. Some people called it an historic coup when Mr. Bezos and Blue Origin LLC successfully launched an unmanned rocket into space and then was able to reuse the same rocket after it floated back to Earth. The significance of the reusable rocket is it could usher in a new era for space tourism within a few years. Jeff Bezos and Blue Origin LLC have been known for their secrecy but after this recent success they felt comfortable enough to release more information. He plans more than a dozen test flights that would demonstrate a capsule carrying passengers can safely disconnect from the first stage in case of an emergency during the ascent. The craft is designed to carry up to six passengers. The company has one location in Seattle, WA with the launch pad located in part of West Texas which is very similar to the Virgin Galactic site in New Mexico. Blue Origin LLC is also developing a new facility at Cape Canaveral, FL.

Since he was a child Jeff Bezos has been interested in space. As a teenager he said he wanted to build hotels in space as well as amusement parks and even colonies that could accommodate millions of people in orbit. The goal was to be able to evacuate humans and preserve Earth. Why he wanted to evacuate humans and preserve Earth, I don't know. As an adult Mr. Bezos has invested more than 500 million dollars of his own money into the project. His net worth is estimated at 46.5 billion dollars. And like Mr. Branson, Mr. Bezos works with NASA and has received funding from that administration.

While space tourism in general seems promising it is currently limited to sub orbital trips where passengers would experience zero gravity but its full potential could be decades away. Still, the interest and subsequent potential is there, especially when considered more than 10 million people each year visit space museums, space camps, rocket launch recovery sites, research and developments facilities and the likes. Space museums and research sites are said to generate about one billion dollars a year to the American economy. With the possibility of zero gravity trips becoming more popular they might be a reality within five years. The emerging space tourism market is on the horizon but there is no way of telling if it is just too expensive to become a real economic boom.

The tourism and travel industry creates a tremendous number of jobs and is the second largest employer in the United States. It generates more than 400 billion dollars a year. But the vast number of tourists and travelers are not the super wealthy. Economists say the base for space tourism would seemingly have to come from the middle class. And therein could lie the secret to a successful space tourism industry. They also say for it is to reach its potential it will most likely have to be made accessible to the masses.

Not only is space tourism expensive but it's also dangerous. There have already been a number of fatal accidents involving at least two of these pioneering companies. The safety factor of the emerging space tourism industry is too low to become acceptable. According to some reports and projections there is a 1 in 100 chance of some sort of fatal accident occurring. That figure may be acceptable for the military but not for a commercial industry which would require that figure to be reduced by at least half. Perception is said to be a major part of reality and if that's case, much of the public might think space tourism is just too risky, believing that NASA and the Russians are the only two powers capable of competent, safe space travel. But at risk of contradiction, another report seems to suggest that the public is more willing to assume these kinds of risks than are the engineers who work on those projects.

Either way some economists say the space tourism industry could be developed into a 20 billion-dollar a year industry in a few decades. The three men that seem to be leading the way are Jeff Bezos, Richard Branson and Elon Musk, arguably in that order. One of their plans seems to be heading in the right cost reduction direction with each proposed flight carrying a pilot and passenger for a sub orbital voyage at about a 100,000 dollars per seat, with two flights taking off and landing daily, again seemingly only for the super-rich. Requirements for the first new passengers could be that travelers must be wealthy and healthy, and in that order.

Can the space tourism industry be compared to the automotive industry? In the late 1800s who knew what lay ahead as the first car was about to be invented. According to historians that year was 1885.

It was a three wheeled vehicle that was constructed in Europe. A few years later the Ford Motor Company was founded in 1903. Still who knew what was coming. It was an economic explosion, looming just a generation away, that would literally change the world. In 1903 the assembly line was perfected by the man who founded Ford. A Model T automobile could be made in a matter of minutes instead of several hours. As a result, the public was initially amazed at the automobile, viewing it as some sort of mechanical marvel from the future. But the moment this mechanical marvel from the future became mass produced and accessible to the masses, was the moment that the public embraced it whole wholeheartedly as an essential part of everyday life.

Could history be waiting to repeat itself with regard to space tourism? This could the make or break point for men like Bezos, Branson and Musk. It seems to be clear there are the wealthy whom are more than willing to risk life, limb and a small fortune to travel to space. But what about all the others who are not wealthy and cannot afford to sacrifice their child's college tuition for a two hour ride on a sub orbital trip above Earth. What are they to do? Henry Ford was able to solve this problem in an inclusionary manner that did not limit his invention to only those who could afford it and in doing so he strengthened society in unimaginable ways. This ethical and inclusionary approach to the automobile became a huge success. Can the space tourism pioneers follow suit?

What is certain is the speed of technological advancements and the changes in society are breath taking. What was thought impossible or unreasonable 100 years ago is not even considered today. The race to the moon is just one of many examples. Many people who witnessed that lunar landing were old enough to remember a time when they traveled on horseback and the automobile was just an abstract concept. What does that promise to people in the present generation, wondering about what space tourism might have in store for them?

What a Marvelous Instrument!

Without the telescope we would not know much about space. We certainly cannot travel to the other planets, stars, solar systems and galaxies to learn about them, yet our knowledge concerning these matters is impressive and extensive. The telescope gets the vast majority of credit. It has identified thousands of celestial images over hundreds of years which have been set out in a linear form. This provides us with our current understanding of the universe. The invention allowed us to learn about the stars from a distance, as passive observers since it became painfully obvious after the Apollo missions that humans could not travel significant distances in space to make these direct observations. Other scientists such as a biologist or a geologist are not hindered by great space and time, they can make direct observations about the majority of subjects they study.

So the telescope is directly responsible for our current understanding about astronomy. It was first patented by a Dutch eyeglass maker named Hans Lippershey. It is unclear what his intentions were because his areas of expertise were optometry and physics. But in 1608 Hans Lippershey patented the design after watching children in a store hold up two glass lenses that made an object in the distance appear closer. Other people argue that he stole the idea from another eyeglass maker who also lived in the same town.

Like many inventions before and after, the telescope would go on to open new and exciting doors that would yield unexpected and important scientific knowledge. The first person to point the telescope toward the heavens was Galileo Galilei. He did so in 1609, just one year after the telescope was patented. Galileo Galilei was an Italian scientist who studied physics and questioned the work of earlier scientists such as Aristotle regarding heavier objects fall faster than do light ones. Galileo Galilei also learned about the spy glass and began to grind and polish his own lenses. Soon he turned his device toward the night sky. He was the first person to see craters on the moon. He discovered Sun spots and tracked the patterns of Venus. He is also credited with laying the foundation for modern astronomy. So at the earliest, more than 400 years ago humans we're learning about moons, planets and stars from a great distance as passive observers, all thanks to the telescope which built the foundation of knowledge that has advanced science to the point it is today.

Isaac Newton was born the year after Galileo Galilei died. Many prominent people in the scientific community consider Newton to be the smartest man and or the greatest scientist who ever lived. Newton was born in 1643 and by his mid-twenties had an understanding of the world that transcended his age and the time in which he lived, the inventor of calculus and the father of classical mechanics he was widely believed to have died a virgin, possibly too engrossed in his work for anything else. Newton's work extended well beyond gravity as is evidenced by the Newtonian or reflecting telescope. Unlike earlier telescopes his telescope focused on how light reflects from the sun and stars through the lenses of this new viewing device. That opened the door to a better understanding of the spectrum of light. A detailed account of the telescope and the people who used it to advance astronomy is too long to deliver its just due. So let's fast forward to the current state of affairs.

At this point there are no prizes for guessing why we know so much about the universe, including its behavior and age. For that we thank the Hubble Space Telescope. Before its launch in 1990 we had not found a single planet outside our solar system. We have now identified more than 400 and the number is growing. Before Hubble, we had no idea the Milky Way contains a billion or more stars. Again we can't go there to find this out for ourselves.

The notion of dark matter, said to make up much of the universe, well once again thank the Hubble for shining light on dark matter. From how planets are created to how black holes behave and everything in between, all can be traced back to the Hubble. It is truly one of the technological wonders of the modern world, one of the most important scientific instruments ever made. It was launched in 1990 and since has recorded images of more than 30,000 objects in space and has taken more than a half a million pictures. Its orbit is some 350 miles above Earth and is named after Edwin Hubble.

Edwin Hubble was the American astronomer who was born in 1889 and died in 1953. To say he played a pivotal role in the advancement of astronomy is an understatement. He proved that nebula, giant clouds of gas and dust where stars are believed to be created, are too far away to be part of the Milky Way. His work changed the way we understood the universe. He also classified entire galaxies and was said to be one of the few people on the same plain as Albert Einstein.

Edwin Hubble died some 40 years before the launch of what was first called the Large Space Telescope which was later changed to bear his name. At the time it was the first of its kind and like Edwin Hubble it has tremendously enriched our lives. Since the early days the Hubble has undergone several upgrades and is now 100 times more powerful than when it first went into space. No doubt, Hubble's scientific contribution is invaluable but it's worth does not stop there. Think about all of the lives that have been enriched by the Hubble. Remember the results of Hubble's work enters schools, homes, museums, seminars and where ever else there is a quest for knowledge. Hubble allows its wide and growing audience on Earth to become explorers in the universe, each and every one of us.

Yes, we are still passive explorers but that's almost immaterial. What's important is that we are privy to the knowledge, without which our lives and imaginations would not be as rich nor as informed. And who thought all this could come from passive observation! In some cases, be glad we observe the results of Hubble's work from a passive observation because it is from a safe distance. Put another way, who would want to be a direct observer to something that would instantly crush or incinerate us, such as a black hole or an exploding star.

Chalk up another one for the Hubble Space Telescope, its latest discovery, the oldest galaxy ever located was unveiled to the press in the winter of 2016. The galaxy is called GN-z11 and was spotted by the light is gives off, thought to be 13.4 billion old. It is the commonly held belief that looking into space can be like looking back into time, so when it comes to this galaxy we are told it is like looking back 13.4 billion years ago when this galaxy was very young. Scientists say the galaxy was created some 400 million years after the Big Bang. On average most galaxies are billions of years old, so by galactic standards GN-z11 was very young when it sent light out to us all those years ago. It is also thought that GN-z11 was one of the first galaxies to be formed, prompting astrophysicists to call it one of the Holy Grails of the discipline but objects like GN-z11 might be so old that they have long since died by the time we see them. GN-z11 could have been seen at the time when the universe was less than 5 percent of its current age. Again we know this because of the telescope.

Is it possible for a space telescope to actually see light from the Big Bang itself? According to scientists no. Why not if we are to believe that looking into space is like looking back into time, then why can't we see all the way back to the Big Bang? Problem is there was nothing to see back then. Before the explosion there was nothing, no mass, no light, no photons, no anything, so there wasn't anything to see.

All good things must come to an end and so must the Hubble. NASA states that time could come as soon as 2018. As time passes Hubble's instruments and components will probably break down more often, eventually past the point of repair. It has been going strong for some 25 years now and has passed everyone's expectations. But when it does leave it might not really be gone, not when you consider Hubble's contributions passed on from generation to generation.

Kepler tries to Answer Are We Alone?

The Kepler is a telescope and space observatory, launched by NASA in 2007 to discover planets like Earth that might harbor life. Put another way, Kepler's mission statement is to answer the big question, are we alone. It cost 600 million dollars and was originally expected to last just one year, but years later it would still be on the job.

It was named after astronomer Jonas Kepler. The telescope and space observatory were designed to look at planets that orbit stars similar to our sun but are outside of our solar system. This is believed the best way to find life, if it is out there. These planets would orbit their stars in what is called the "habitable zones", not too far and not too close, but just the right distances from their stars.

At one point Kepler was about 75 million miles from Earth and its speed changes depending on its distance from the Sun. It has the largest camera ever launched into space and its telescope is so powerful that from nearly a 100,000 miles up it can detect a person on Earth turning on and off a porch light. It is hard to imagine looking across the entire United States to see that person on the porch, let alone looking nearly 50 times that distance to see something so small and far away with such clarity.

A closer look at the technique Kepler uses to detect the possibility of life is long and tedious. It looks at the big picture, viewing more than 100 thousand stars and tries to identify a small decrease in the brightness in light which can indicate a planet has crossed in front of its host star. Its looking near the Milky Way between the constellations of Lyra and Cygnus, thought to be more than 10 trillion years away.

Despite its huge price tag its design is somewhat simple which includes a large digital camera and some light weight mirrors. In fact, its digital detectors are just like the ones in our cell phones. In this case bigger is better, because a bigger camera and mirrors are more sensitive. It has not found any signs of life yet. But Kepler has found some planets similar to Earth but some are too big for habitation as their tremendous gravities might crush us. Other planets are made of gas with no solid surface.

Kepler was the first telescope to find a planet the size of Earth in a habitable zone along with two planets that have oceans. Again the problem is distance and time because those planets are about 1,200 light years away. As of January 2015 Kepler was still on the job and this is despite a number of problems and break downs that forced engineers to modify its mission in order to keep producing information and collecting data. At last check it has identified 1,013 planets and 440 stellar systems, not mention another 3,000 objects that might be planets. We'll learn more about the Kepler later in the book.

The conversation about telescopes would not be complete without mentioning the James Webb Space telescope. It is called an icon, a triumph of engineering and possibly the most famous craft of its kind in decades. It is named after a NASA administrator who worked on the Apollo project. The James Webb Space telescope resulted from a collaboration of American, European and Canadian efforts and it is big, the size of a tennis court. So big in fact that it will have to be folded for the flight and will unfold once in orbit. While the Hubble orbits Earth at an altitude of some 380 miles, the Webb will soar to a height of about a million miles up, well beyond the Moon's orbit, about four times farther from Earth than is the Moon. Its huge 18 mirrors are coated with a thin layer made of 24 karat gold. Each mirror carries enough gold to make up a golf ball. Gold was chosen for its reflective qualities, making its mirrors 98 percent reflective as opposed to the 85 percent margin achieved by regular mirrors. The Webb can detect infrared light that comes from the farthest objects in the universe. Apparently everything in existence puts out infrared radiation. It is so powerful that it could see a penny from a distance of 24 miles or a football from a distance of 340 miles. It is said to be seven times more powerful than the Hubble and is expected to detect water on planets that are outside of our solar system. The sunny side of its mirrors reach 185 degrees and the shaded side goes down to about 380 degrees below zero.

The Webb also operates in the extreme cold of interstellar space where the temperature can drop to about 450 degrees below zero. It is also expected to see light from some of the first stars in the universe that were formed a few hundred million years after the Big Bang. Planning for the telescopic marvel began it 1995, possibly getting its motivation from the Hubble which was launched five years earlier. The Webb's launch is scheduled in 2018.

Our Natural Satellite!

Ever wonder what life would be like without the moon! We're told there would be none, or at least no advanced life as we know it. Here's the story of our moon according to the latest theory. We're told the moon was created by Theia. Theia was the Mars sized planet that crashed into Earth around four billion years ago. The crash was not direct nor high speed by galactic standards. Instead it was a little more than a glancing blow, but had hit directly both planets might have been destroyed.

Before that glancing blow Earth might have rotated much faster than it does today. At that point a day on Earth could have lasted about eight hours or so. Shortly after the blow Earth absorbed most of Theia into its core with the rest of the debris getting shot out into orbit. In a matter of days, the debris from Theia began to coalesce and gather into a larger body around Earth which would form the Moon. We know material like loose rocks in space will be attracted to the gravity of other larger loose rocks in space and will eventually create one big rock. So WA-la that's the theory on how the moon was created.

At first the Moon was much closer to Earth than it is today, at which point it would have taken up the entire night sky. Can you imagine that full moon! The moon's gravitational pull on the oceans was also much greater at that time, creating tides that might have swept hundreds of miles inland and then rushed hundreds of miles back out, twice a day. The great tides swept all sorts of minerals, salts and other nutrients in tremendous amounts, right out of the soil and deposited them into the ocean. This is how the oceans became the rich, salty breeding grounds for life that they are today. The Moon also created a stabilizing effect on the Earth.

Immediately after the impact, the Earth was tilted on its axis and still is today. The Earth wobbled so wildly on its axis that life as we know it could not have developed. Again the tides and storms in this condition were just too severe for the development of advanced life. But in a short period the Moon coalesced from the debris left behind from Theia and stabilized Earth so our planet no longer wobbled on its axis. The Moon was securely in orbit and continued to move away from Earth, much like a ball swinging in a circle at the end of a rope. The Moon continues to move away from Earth at the rate of about an inch and a half a year. It will eventually move far enough away as to be clear of Earth's gravity. But because the Moon was initially so close to Earth, the Moon became tidally locked, meaning the Moon is so close to Earth that the Moon does not spin on its axis but instead always shows us the same face. This phenomenon gives new mystery to the old saying "The Dark Side of the Moon."

But why does the Moon appear to be the same size of the Sun? The answer might be lie in the acronym PAD, short for perspective, alignment and distance. The Moon is much smaller than the Sun, in fact you could fit 400 Moons inside the diameter of one Sun. But the Sun is also 400 times farther away than the Moon which is why the Moon and Sun appear to be the same size when viewed from Earth. Put another way the Moon is about 400 times closer to Earth than the Sun which is why the Moon looks so much bigger and the Sun so much smaller.

The Sun and the Moon are also aligned, meaning they are on the same plain with one lining up one in back of the other at certain times which allows us to compare their sizes. Moons often orbit their planets at the planet's center of gravity which is at their equators and not at their polls. Makes sense when considered there is more mass to generate more gravity at the larger equators than at the smaller polls. This is also common when planets orbit stars. So this explains their alignment, which again allows us to compare their sizes. And because there is nothing in the sky to add perspective to the Moon and Sun they appear to be the same size when they line up.

Scientists are trying to figure out the origin of the moon's magnetic field. The fact the moon even has one comes as a surprise to scientists who explained what is needed for Earth to have a magnetic field and apparently the moon lacks the basic requirements. The Earth has a hot liquid metal core and an atmosphere and those are some of the basic requirements. The moon has neither. But how are some sides of the moon magnetize? Scientists theorize that an asteroid about 120 miles wide slammed into the moon's south pole some 4.5 million years ago and left behind the necessary metals for magnetism. Other scientists think the moon's magnetism was caused by a series of smaller asteroid impacts. As one astrophysics sarcastically said, what good is the moon's magnetic field when there is no life to protect?

But does a full moon really influence human behavior? Let's examine the record. Remember the human body is about 75 percent water. We know the stronger effect a full Moon has on Earth's ocean tides; in fact, a tide is said to be 20 percent higher during a full Moon than at other times. According to one report full moons have no effect on the number of emergency room visits, mistakes made by surgeons in operating rooms, psychiatric visits or even in the reporting of epileptic seizures.

But there are reports that indicate pet injuries increase during periods of full moons by nearly 30 percent. There is speculation that this could be because an animal's sense of awareness is more acute than humans. However, there is no definitive proof that animals become more aggressive to humans which would result in an increase in the number of bites during full moon periods. And what about a woman's menstrual cycle? A full moon Occurs once a month. A woman's menstrual also occurs once a month. But that seems to be the only connection with scientists reporting one does not lead to the other.

The Free Light Show!

The Aurora Borealis, known as the northern lights have amazed and confused humans for a millennium. Early humans once thought they were some sort of sign from the gods or maybe the edge of the world or even the reflection from a large fire. The light shows have been depicted in paintings on the walls of French caves dating back some 30,000 years. Aristotle called them jumping goats. Galileo described them as the dawn of the north and Native Americans thought they were gods dancing in the sky.

Thanks to modern science we know the lights are a demonstration of Earth's magnetic field as it interacts with charged particles from the Sun. The Auroras are centered on our planet's magnetic poles. But the geographic and magnetic poles are not the same, meaning the lights can be seen farther south than one might think. These lights also shine over the south pole in Antarctica but that area could be too sparsely populated to report dependable viewing.

There are reports the lights were seen as far south as Fredericksburg, Virginia during the Civil War. The reports came from the diaries of several soldiers. One possible explanation for a siting so far south is, at that time there was no population explosion with large cities and bright lights. Bright lights can of course wash out Auroras. If the Auroras occurred over New York City you could look up and not see them. But in modern times the lights were seen as far south as Florida and Cuba which recorded siting's in 1989 during a geomagnetic storm.

The interaction between Earth's magnetic field and the charged particles from the Sun takes place about 62 miles above the Earth's surface, again the commonly accepted altitude at which space begins.

The Sun's charged particles include protons and electrons and they slam into Earth's atmosphere at high speeds and that's when they release their energy which produces light. That light can be seen in different colors depending on the makeup of the particles. For instance, oxygen atoms produce green lights but are also sometimes red. As a general rule the green lights are seen higher in the atmosphere and the blue or purple colors which can indicate nitrogen are seen in the lower elevations. Because the lights from the Aurora Borealis are dim they can be better seen with a camera than with a naked eye. Cameras are more sensitive than a human retina. Even though the northern lights look like they are on fire, they are not, they are really cold, around 40 degrees below zero.

There are no prizes for guessing what the northern lights are called in Antarctica, they're called the southern lights or Aurora Australis. There are certain times of the year when the lights are most visible, usually from early fall to early spring. Another time when the lights are thought to be most active coincides with the eleven-year cycle of the sun's surface activity. This period is called the solar maximum and it's when there are more sun spots and more flares than normal, which as a result release more powerful bursts of radiation that are headed toward Earth. They are also a boost to the tourism industry with folks all over the northern hemispheres scheduling vacations around what has been called the greatest light show in the world.

The lights can also be heard. Sounds from the auroras cannot be heard during every light show but they do emit sounds that are faint and brief. The sounds are said to be like strange clapping noises. Some were recorded some 230 feet above the ground and others almost a mile up. But the lights are also dangerous. They convey solar radiation and at one point Apollo astronauts could have been exposed to lethal amounts of the radiation. Scientists at the International Space Station are prepared to modify their activities in order to avoid exposure to high levels of radiation during these temporary episodes. So the Aurora's exemplify one of the many times when beauty and danger go hand in hand.

My How Time Flies!

It is hard to believe that we've been on the surface of Mars for more than 40 years now. It was 1975 when project Viking blasted off from Cape Canaveral, FL and touched down on the Martian surface one year later. There was a machine on board that would go on to make history by sending the first pictures of the Martian surface to Earth. That first famous shot was called Face on Mars. Trips to the Red Planet also confirmed the theory of Gravitational Time Dilation. In short, this means times passes different when it goes through different gravities, or more precisely time will pass at different rates when it is exposed to different gravitational forces, which was proven by using radio signals.

Traces of water in glaciers and close to the surface have been found on the Red Planet and rocks there seem to be similar to rocks on Earth which should come as no surprise because the two planets are alike. A Martian day is about 24 hours and 37 minutes. Its year is 686 Earth days again very similar. But gravity on Mars is a bit different than on Earth, it's about one third less so if you weighed 110 pounds on Earth you'd weigh about 38 pounds on Mars.

Considering all that we've learned from Mars over the last 40 years many people don't advocate trying to send humans there to learn more. For starters the trip would be almost cost prohibitive. It costs about 10,000 dollars to send a pound of cargo into orbit. Mars at its closest is about 33 million miles from Earth and about 250 million miles away at it farthest. So it might take at least six months to get there when Mars is at its nearest point, which only occurs about every two years because of its orbit. Here's just some of what we already know about Mars. It could have been home to ancient life. Our most recent Mars rover found some of the ingredients that are required for life as we know it. There is evidence of water flows found in ice and in rock below the surface. We know about the atmosphere and gravity of Mars. Dangerous levels of radiation could also be present. There is a surprising amount of geological diversity on Mars such as volcanic rock, mud stones and cracks containing minerals but there is no methane and no life.

Scientists say humans were not meant to spend much time in space. As mentioned earlier, our bodies begin to deteriorate immediately when exposed to weightlessness, such as muscle and bone marrow loss. If someone wanted to spend at least six months trying to get to Mars, they could only spend a few weeks once they got there. That's when the two planets would be closest to each other in their orbits. If too much time was spent on Mars astronauts might never survive the several million-mile journey home. The trip to Mars is so long that two space craft might have to be sent, one for cargo and the other for the crew which could double the cost of the trip. Opponents of a Martian voyage say humans could only spend a few weeks on Mars after braving the dangers of at least six months in space just to get there. It is highly doubtful they would have enough time to contribute to the substantial base of Martian knowledge that already exists. In short opponents say the trip is not worth the time, money nor risk to life considering what little new information might be collected from the cost prohibitive journey that would span at least 66 million miles and take up to two years to complete. Later in the book we'll review the results of a new study suggesting Mars had an ocean and might have been hit by asteroids that caused tsunamis.

Creating a Home!

Having given ample time to people who are against humans traveling to Mars, it's only appropriate we briefly examine proponents who likes the idea. Human travel to Mars might be able to lay the ground work for something called terraform which at this point is complete science fiction. But so was lunar travel until it became reality. Although the difficulties involved with terraform are much greater than traveling to the moon.

Terraforming is the transformation of an alien planet into something humans could call home. A writer came up with the term in 1942 for a science fiction story. At its most basic the idea involves heating a planet and altering its atmosphere, kind of like what we've done on Earth, but the planet must first be uninhabitable, unlike Earth. We might have the capability to alter some elements of a planet but at present, the economic resources required are far beyond our means and that is not to mention the time element. From start to finish terraforming might take at 500 years or longer.

Mars being the nearest planet is the most likely choice for such a concept. Below is one of many terraform ideas on changing the Red Planet. The first human mission to Mars would lay the foundation for a lengthy project that could span several generations and maybe last a thousand years. The first human crew to reach Mars would construct some sort of habitat for the next crew to build upon. The second crew would do the same for the third generation and the third would do the same for the fourth. This linear behavior would continue for several hundred years until bit by bit there was a large enough structure on the Martian surface to call home. Again the structure would be too large to transport in one or two trips so it would have to be transported and constructed in the aforementioned stages.

That habitat would most likely be some sort of inflatable pod from technology that already exists. As you can imagine it is built of durable, tough and flexible materials that could accommodate six people. There is a company based in Nevada that can already manufacture these materials and its founder built the company's fortune from the hotel chain Budget Suites of America. So part of the private aerospace industry is already here even if terraforming is still just a dream.

Next, humans could turn their collective attention to transforming Mars' atmosphere. This would entail thawing out parts of the Martian soil in an effort to release carbon dioxide trapped beneath the ground and inside its polls. Some sort of reflective devices might resemble huge mirrors that could speed up the process by shining ample sunlight onto ice. This might be one way to heat the planet which would be required because the temperature at the equator of Mars can drop to about 100 degrees below zero. Equators are often the warmest regions of a planet.

Advocates believe that much of the carbon dioxide that once warmed Mars is still there but is now frozen. They also believe much of the water that once flowed on Mars is still there as well. The hope is that the planet can be warmed to the point where it would boost the pressure in the atmosphere. It would then be hoped that the atmospheric pressure boost would increase the heat and make water that is frozen underground flow to the surface. In turn the process would begin to make the Martian environment somewhat habitable for humans. The operative words are "somewhat habitable" which will be explained in the next paragraph.

The process of evolution would start to feed on its self but it would not be a natural development like what occurred on Earth. Instead the evolution on Mars would of course be assisted by humans. One might expect to first see lichens and then mosses. Maybe after 500 years or so the first plants and trees would start to grow. Many years later these plants and trees could open the door for oxygen levels that might actually be breathable. Advocates are quick to point out that for the foreseeable future life on Mars would be like living on the south pole in Antarctica where temperatures can drop to 115 degrees below zero.

The dream of turning the Red Planet green is still just that. Considerations such as time, money, reality and the lack of technology are formidable obstacles to overcome. Those against the idea can come up with many more reasons why the terraforming of Mars should stay in the realm of science fiction.

So what does the American public think about traveling to Mars, the idea of terraforming in particular and space exploration in general? Reactions are mixed with a large percentage of those surveyed backing space exploration but with a smaller percentage not wanting to fund it, at least with their tax dollars. According to one survey, 58 percent of those polled said its essential that Americans be the world leaders in this endeavor. Little more than a third of the people surveyed said they like the idea of space exploration conducted in the name of science, instead of something more sinister such as using space exploration for warfare. But they also say that money could be better spent on the solving the problems of everyday life on Earth. However, a majority of those polled say space exploration does more good than harm.

When it comes to the International Space Station more than half of those surveyed said it was money well spent, while a third of the people polled said it was not. About half like the Shuttle program and said that it was a good national investment with about a third saying it was not money well invested. In general Americans seem to agree on the value of space exploration but disagree on its importance, which can be confusing until you realize something can be of value but not be important. For instance, a diamond might be of value but has no importance. And despite its successes and failures, a majority of Americans have a favorable view of NASA and they remain optimistic as to what the future might hold. The common denominator throughout the survey was money, with most Americans being reluctant to spend more of it, especially tax dollars on space exploration which doesn't seem to bode well for such an expensive concept as terraforming.

Who's Out There?

If the alien character is anything like the human character, then we don't want to meet them. One of the most talented experts on the subject has espoused that opinion for more than a decade. Well known Mathematician Stephen Hawking shared that opinion with the public in 2010. In short he said an advanced alien civilization might wipe out the human race in the same way humans would wipe out a colony of ants. If were visited by aliens they certainly would be more advance than humans by the very fact that they know how to travel to distant planets and we don't. In such a scenario our only hope might be that the alien character would somehow be different than ours. But Mr. Hawking says why would they be.

Examine the record. Humans have behaved in the worst way imaginable toward other people since the beginning of recorded history. We have not hesitated to brutally wipe out those who are less technologically advanced in the name of greed, racism and religion and in the process we took from the victims whatever they had of value. Mr. Hawking suggests aliens could take Earth and all of its resources for whatever needs they deemed suitable. The aliens might be nomadic, intending to capture, conquer and colonize, which would mirror human behavior. If they were like this it would only stand to reason that they would try to exploit each new planet for its raw materials that could be used to repair old space ships, build new ones and then move on once Earth's resources have been exhausted. Humans might be exterminated in the process. Who knows what intelligent aliens might be capable of! If human behavior is any indication, we don't want to know.

But does curiosity trump fear? We know it killed the cat but will it kill humanity? Again, you know what Stephen Hawking has to say. But apparently his curiosity is stronger than his fear because he recently announced his participation with an initiative called Breakthrough Listen. The Breakthrough Initiative is a program designed to search for intelligent alien life and it is broken down in two parts.

Breakthrough Listen focuses on searching millions of planets and stars for artificial radio or laser signals. Breakthrough Message focuses on creating a message that would represent Earth and humanity. Both were announced in the summer of 2015. It is a search that might yield the best of results or the worst. Of course scientists have no idea what alien life might look like, where and when it might be found and how it might react to us. It is a tricky, high stakes proposition.

But other theorists say intelligent alien life may already know we exist and are waiting to see if we evolve to a higher state before reaching out. Or stated more directly, they could be waiting to see if we become better people, capable of contributing more than space junk to the solar system.

The Search for Extraterrestrial Intelligence or SETI is one of the more recognized and established efforts in trying to answer the age old question of are we alone. SETI has monitored electromagnetic radiation for signs that intelligent life on another planet might have already sent out some sort of message. Radio signals also penetrate Earth's atmosphere and have spawned the construction of large radio antennas. But so far nothing. In fact, in the absence of fact some scientists are wondering how could this be, with hundreds of millions of stars and planets that could offer millions of chances of finding intelligent life, yet we hear and see nothing. Maybe we have not been looking long enough or maybe we're looking in the wrong direction. In the late 1890's an American inventor thought he could use his electrical communication device to contact beings on Mars. We now know Mars doesn't have any. Are we close to coming to the same conclusion about the solar system?

In the search for intelligent alien life we seem to replace fact with optimism, maybe these aliens have already tried to contact us and we didn't recognize the message. Maybe the alien communication was just too advance to understand? Take a gamma ray burst for example. They are super high energy bursts of light, heat and radiation. They are detected about once a day. Their frequencies have been monitored by humans are said to be able to contain and transmit tremendous amounts of information in one millionth of a second. Intelligent life could have already sent us a message equal to all of the knowledge acquired by humanity and we might have missed it in the blink of an eye. But other scientists say if messages have been sent we should be able to detect at least a part of them. The only thing we know for sure is that we have not heard from anyone or anything but that's not from a lack of trying.

Any search for life outside the planet should probably include a conversation about viruses and their possibilities. Scientists tell us a virus is something not quite alive but not quite dead and instead is somewhere in between. A virus is too small to be seen with an electron microscope and is really kind of a strand of DNA. It cannot provide its own energy nor can it replicate itself. A virus survives by infecting other cells which supplies it with the required energy to grow and reproduce. Could viruses be on other planets? If the answer is yes, did these viruses come to Earth from asteroids or comets? What is known is that viruses evolve from a non-living organism into one that has life through its parasitic behavior. Do we need to expand our definition of life to include something like a virus once it has found a host? And if so, did viruses come to us from another planet instead of originating here on Earth? If they did could viruses be considered alien life that has adapted to Earth?

And in this search for alien life are scientists too quick to eliminate planets whose environments seem too hostile to host life? And could this be a possible reason why we have not found any life? If a planet does not have the requisite amount of sun light for moderate temperatures or too much surface pressure it is likely discounted as inhospitable. But there is a place on Earth just as inhospitable but life has taken hold, in fact entire food chains have prospered and proliferated. It is called the Marianas Trench possible the deepest ridge on the planet, located several miles underwater. The trench does not see sun light, it has eight tons of pressure per inch and is home to food chains and multiple species. This underwater world on Earth is every bit as alien as one the ones we theorize about in space. Life in the trench is not indisputable proof that life in space can take hold in extreme conditions but it does suggest that it can happen because it has happened in the Pacific Ocean where the trench is located. It is more than 1,500 miles long and only 43 miles wide considered the deepest point on the planet. Down there the pressure is 1000 times greater than at sea level, temperatures too cold for humans to bear but yet dozens of species, big and small thrive seven miles down in complete darkness. The lesson from the Marianas Trench is we could have already looked in the right spot for alien life but discounted it as uninhabitable and in the process passed over life that was waiting to be discovered.

UFOs!

When I was younger I thought believing in UFOs was the same thing as believing in alien life but now know the two are completely different. A UFO could be anything that is not identified and happens to fly. Alien life is much more specific but yet more broad at the same time. UFOs have been seen before, aliens have not. But it is still hard to dispute the extensive and growing evidence regarding UFOs. They would be easier to dismiss if there were just a few sightings by a few individuals, but down through the years there have been many sighting's, many by credible sources and they've also been viewed by large groups of people at one setting.

Intelligent alien life has never been documented. There has been no recording in the form of hard evidence that can be re-examined by several independent objective sources. Who knows what UFOs actually are, maybe military craft, weather balloons, experimental flying saucers, it's anyone's guess; but they have been the source of intense discussion and spawned numerous stories, movies and rumors. Ultimately these sightings ask the age old question, are we alone!

UFO sightings have been observed almost since the beginning of recorded time from the early Romans to modern humans. One of the first purported sightings was recorded in the year 214 BC and they continue to come in from all over the world. In 214 BC observers thought they saw an ocean going ship that was shining in the sky. Current observers reported seeing lights moving at astonishing speeds and in directions not capable by known aircraft. In 2008 a police officer reported seeing a UFO as he drove on the highway. He reported seeing orange lights in a strange configuration in an area that was known for bizarre sightings. Another one was seen over a college campus in Tennessee in 1853 and then there was the sighting in 1942 over Los Angeles. Radar operators reported seeing that one and tracked it until it just disappeared from the screen somewhere over San Diego. In the summer of 2001 another one was seen above the New Jersey Turnpike and was witnessed by many motorists. In 2008 there was a reported UFO called the Stephenville Lights and was said to be the size of a Walmart. They could have called that one the Blue Light Special.

Next are some of the more extraordinary sightings. In the middle part of the last century many people reported seeing a green fire ball race across the sky in New Mexico. In these reports the UFO was seen by just about everyone in the vicinity including military personnel. Many observers thought they were seeing asteroids, including government personnel who brought in an expert on the matter to provide answers. That expert spent years studying the phenomenon and decided that the lights were too slow to be meteors and there was no evidence in the form of rocks or debris left behind. Their conclusion was inconclusive. The Air Force stated the lights were caused by some sort of sun spots. The expert that was hired to investigate the matter said that statement did not make sense. In fact, many national militarizes have detailed files about these reported sightings.

Another noteworthy sighting came from the British Military in 1952. In this case a combat pilot reported seeing a silver flying saucer that appeared out of nowhere and flew next to his jet for quite a distance. The saucer reportedly intercepted the jet but it could not be determined if the saucer's intent was based on curiosity or something more sinister. It finally flew off without further incident. Radar on the ground also tracked the saucer and reported that it traveled at speeds that were far greater that any known aircraft. Even though the military has been known to downplay UFO sightings, it called this one "convincing." The American military reported the almost exact sightings on US soil days before and it concluded that the events were related. But one military official called the sightings "mass psychology" but after he was promoted and pressed on the issue he apparently retracted his bold statement. In the follow up announcement he basically said, that in the absence of credible scientific evidence that could provide an explanation, these sightings in general and that flying saucer in particular cannot be characterized as a "mild form of hysteria." And even though that British combat pilot who originally saw that flying saucer died in 2009 he passed all of that knowledge on to his son which was intended to be documented in a book.

In 1986, passengers and pilots on a Japan Airlines flight over Alaska reported seeing a bright object outside their windows that was about three times the size of an aircraft carrier. The average carrier is a thousand feet long and weighs about 100,000 tons. That would make it about three times the size of a football field. The pilot contacted ground control which confirmed the object on radar. They tracked the UFO for more than 300 miles before losing site of it.

Almost ten years later in Phoenix, AZ residents reported seeing a series of bright lights that were actually seen over five cities in that state. The lights were configured in a V shape pattern and were visible for about four hours. At the 2012 Olympic Games in London a UFO was spotted during the opening ceremonies in what appeared to be behind a fireworks display. Those who doubted the authenticity of a UFO sighting said it was probably just a blimp or helicopter but organizers said no such objects were supposed to be in the area at that time.

Part of the problem with the UFO phenomenon could be rooted in the pace at which technology has evolved over the last 50 years. We've been to the moon, sent machines to other planets and made tremendous advances in medicine that have prolonged our collective lives. This progress has spawned an attitude that anything is possible, including UFOs, which may or may not be true.

Reported sightings of UFOs don't answer the question of are we alone, but they suggest that the answer would be no. Is it more probable than not that UFOs are flown by intelligent aliens as opposed to mischievous humans? Again UFO sightings have been coming in almost since the beginning of recorded history and in that time there is no evidence, not a single shred of proof that would confirm the existence of intelligent alien life. One theory is that intelligent alien life should look similar to humans because evolution is fairly predictable. So if they are out there we should be able to recognize them. But again there is no evidence of their existence, just theory and sightings. In the absence of evidence one could come to the conclusion that we are alone or at the very least reserve judgment. On the other hand, the absence of evidence could in fact be evidence that there's no one else out there.

But what about the law of averages! Consider the universe, said to contain some 100 billion galaxies most of which are much older than the Milky Way which in theory would give intelligent alien life a head start on evolution, but yet there is no sign of anything. Maybe they are too far away and despite their advanced technology they still cannot traverse the great distances to contact us. All of the relevant, factual data strongly suggest that there is no one out there but us. How strange would that be, life in the universe is rare and precious but on Earth it is plentiful and at times worthless. Of all the strange phenomenon in the universe, the proposition that life might only exist on Earth could be the strangest of all.

Dark Man's Dark Star!

A black hole is an area in space where gravity is so strong nothing can escape, not even light. We are told black holes consume anything that comes too close, asteroids, planets, stars and even galaxies. The one at the center of our galaxy is predicted to eventually consume the Milky Way from the inside out. As we also know, black holes are said to be created when a massive star collapses in on itself.

But who and how was it discovered? How appropriate that a black hole was discovered by a Black man. The idea was first put forth in 1783 by John Mitchell. There is no known picture of him. He is described only as "a little short man, of black complexion, and fat." That may not be the most flattering of descriptions and very little is known of John Mitchell to this day. That could be due to the racial temperament of the time or because Mr. Mitchell did very little to promote himself or his ideas or a combination of the two. But this dark man envisioned a dark star that we call a black hole today.

He was born is 1724 and was considered one of the most brilliant scientists of his time. He was educated at Cambridge University and taught Hebrew, Greek, mathematics and geology. He entertained many visitors who admired his work including Benjamin Franklin as well as other great thinkers, scientists and inventors who saw Mitchell's production as both enlightening and inspiring.

John Mitchell's work on black holes was conducted in stages over a period that lasted more than 30 years. Some of his earlier work focused on the magnetic forces exerted by each pole of a magnet and how those forces decreased with distance. After a huge earth quake of that day, Mr. Mitchell was credited with helping to establish the science of seismology. He later experimented with an apparatus that could measure gravity's force between two objects, which we now know as mass. Mr. Mitchell would die before his work would come to fruition but the scientists who carried on that work gave Mr. Mitchell full credit. John Mitchell was also the first person to apply mathematics to astronomy and was also the first to imagine the concept of a black hole, which again he called a dark star.

He did this by studying other stars and came up with a hypothetical method for measuring the mass of a particular star. He thought that the speed of particles and light traveling in space would be slowed down by the star's gravity. He then thought, what if a star's gravity was so strong that the star's escape velocity would exceed the speed of light. Escape velocity is defined as the lowest velocity or speed that an object must have in order to escape the gravity of a planet or star. He then reasoned that the light would fall back down into that star. He used his considerable mathematical talents to calculate in order for this to happen that star would have to be really big, at least 500 times the size of our own sun. He concluded that if light cannot escape from such an area, that area would have to be invisible to the outside world which he thought would be a dark star and WA-la the black hole theory was created.

Some two hundred years later, thanks to Einstein's theory of relativity, we now know that light travels at a constant speed and is not slowed down by intense gravity, but John Mitchell got everything else right, two hundred years earlier when he proposed that any object that has an escape velocity stronger than light will be invisible. The dark star concept was so far ahead of its time that it was paid little attention, possibly because people did not know enough at that time to comprehend it, including other scientists.

Fast forward to the present, we now also know that a black hole with the mass of our planet would be about the size of an inch. That's right

Earth would be just one inch. And if a black hole had the mass of the sun it would be just a mile or two across. Mr. Mitchell thought we might be able to better detect these black holes if some of them were surrounded by bright stars. Another theory spawned by the work of Mr. Mitchell was that our entire universe might be a massive black hole tucked inside a much larger universe.

One more theory that was built on the earlier work of John Mitchell is, black holes consume everything that comes too close. But now we know that this might not be true. If a star collapsed into a black hole its planets might remain in the same orbit because the mass of the star has not changed, only the size of the star would change. So with its mass in tack, albeit much smaller, its gravity would also remain unchanged. Again in 1783 most people, even the brilliant scientists of the era were ill equipped to comprehend the amazing work of John Mitchell. But our current understanding of the many black holes known to exist today, their escape velocities and the likes were all built up the shoulders of a black man who was described in less than flattering terms.

If it's a Man's World, then it's a Woman's Galaxy!

The person who lead the way to understanding how the galaxies rotate is a woman. Her name is Vera Rubin. She was a Jewish American born in Philadelphia, PA in 1928 and started studying astronomy at the age of ten. As with most everything, the sooner the better. At 14 she built her first telescope. Later in school she was twice advised against pursuing astronomy, once from a physics teacher in high school and then again as she entered college. We should be glad she was not deterred. One of her first papers got lots of attention, not much of it though was positive but at least she was on the map. Her work continued to cause controversy, especially when she determined that the galaxies were not evenly spaced as the Big Bang theory suggests.

She was a devoted wife and mother and a strong advocate of getting women involved in astronomy and other scientific fields which at the time were reserved for men. In short, she uncovered the way that galaxies move by studying galactic rotational curves which solved the mystery of galactic rotation.

That work would lead to something even bigger, it would shed light on dark matter. Ms. Rubin thought that the galaxies would fly apart if the only thing holding them together was the many stars that were contained inside of them. She thought that galaxies spin so fast that the stars inside of them just don't have enough gravity to hold everything together and wondered why everything does not fly apart.

Her work was based on speed, rotation and gravity. She then thought there must me something else at work and there must be a massive amount of matter, that is unseen, acting as a stabilizer which holds these galaxies together. She then calculated that galaxies must contain at least ten times the amount of mass that can be attributed to its stars. And WA-la the theory of dark matter was born. In 1970 she strengthened her theory by putting forth how dark matter significantly increases the rotation of a galaxy. Vera Rubin's significant accomplishments illustrate the liner nature of astronomy, in that much of what we know today was built upon knowledge from previous scientists and part of Ms. Rubin's discoveries were aided by similar work produced in the 1930s.

She also studied the galaxy nearest us which is Andromeda and learned Andromeda was spinning at a really fast rate but its stars were not flung away. She also found that others galaxies behaved much like Andromeda and the Milky Way. Even though her work was controversial she was not the first astronomer to suggest these theories. But her work was so ground breaking that without it, we would not have our current understanding about dark matter and the universe.

In fact, dark matter and its energy could make up as much as 90 percent of the universe. Her work charted new courses that will dictate the paths that astronomical research will take for years to come. Controversy never seemed to be far away even though her work had become widely accepted and greatly admired. At one point she said "fame is fleeting... my numbers mean more than my name." She also said that the biggest compliment would be that scientists are still referring to her work years in the future. This young Jewish child who loved astronomy would grow up to become a juggernaut that changed the way we view the universe. Just imagine what might have been lost had she taken the advice of two teachers so many years ago who advised her to pursue disciplines that were more appropriate for her gender. And she has achieved her biggest accomplishment because years after her death we are still referring to her work.

Science from Science Fiction!

The man known as the father of rocket science had an uphill climb from the very beginning. He probably had to fight for almost everything he had being the fifth of 18 children. He learned about rocket science with little more than a card to the public library. That's a proper introduction to Konstantin Tsiolkovsky, born in 1857 and at young age he developed scarlet fever which left him deaf. He was already withdrawn but his illness amplified his condition. He was eventually suspended from school but the young Tsiolkovsky was still obsessed with books and apparently would not be denied his destiny.

He also read science fiction, Jules Verne in particular. After he read one of Vern's books he thought if a cannon used in the novel could really launch a rocket into space it would kill everyone on board. But the seeds of imagination were already planted. He turned ideas gathered from science fiction into pure science. I wonder what he would have to say about the concept of "sight speed?"

His father sent him to Moscow where he spent countless hours inside libraries reading about chemistry and physics. College was out of the question because of the cost but he was determined to reach out and make his own future despite his limitations. So he made an ear trumpet by himself and attended as many lectures and studies as possible in the library. He was probably impervious to the looks from others in the pristine library setting as he sat there holding up that big ear horn, its large open end directed at the lecturer and the smaller end attached to the side of his head, snugly into his ear.

His early years were filled with trial and error, in part because he was not formally trained in physics but he still built a contraption that he thought would someday make it into space. It was based on centrifugal force and it stayed on the ground, which meant his academic reputation also stayed on the ground. He then had to go back home and became a People's Teacher which was the lowest category of educational instructor of that time. While teaching the public he continued his physics studies which included publishing a number of papers. One even contained a drawing of a space ship powered by a rocket. Along the way he developed Russia's first wind tunnel. It was not long after that he came up with the equation that involved the exhaust velocity of a rocket and its mass relative to its speed. Today that is known as the basic rocket equation. He also realized that his method must be conducted in steps which today is called staging. As the stages burned out they would drop away and the upper stages gained speed, just like the modern rockets that would come so many years later.

In 1935 he knew he was dying of cancer so he tried to persuade the Russian Government and the Bolshevik Party to provide a pension for his family. His home later became a museum that still bears his name. Besides being known as the father of rocket science he left behind many profound quotes, made even more powerful by the time in which they were proposed. Among some of his most famous: "The Earth is the cradle of mankind, but mankind cannot stay in the cradle forever" as well as" the most important things still lie ahead." He also said that he does not remember when or how he got the idea for the first rocket design but thinks it came from reading science fiction. His words place emphasis on the power of thinking out of the box and the ability to imagine ways in which to make the impossible, possible. Remember many of his ideas were well before their times and came in an era when people still rode horses as a common means of transportation which provides proof that certain dreams can come true.

A discussion about rockets would not be complete without mentioning Robert Goddard, born in 1882. He too was born into a modest childhood and was also inspired by science fiction, this time it was H.G. Wells and the War of the Worlds. After reading the novel he dreamed of building some sort of space craft, possibly inspired by the science fiction story he just finished reading. He would continue to dream about creating the impossible. At one point during his inquisitive childhood he thought he could jump higher if he rubbed his feet in zinc from a battery, which of course didn't work. But his imagination was already full of ideas from those science fiction novels and he went ahead many more experiments as a child but stopped after his mother told him if he succeeded, he just might fly away and never come back.

When he was older he came up with the idea of traveling along the east coast at record speeds in a vacuum tube devoid of air pressure. The mode of travel would be by some sort of magnetic attraction that would pull instead of push a vehicle. He envisioned a train in a vacuum tube that would make it from New York to Boston in ten minutes. 120 years later scientists were still pondering the idea but this time with a vacuum tube type train called the Hyperloop that could travel from San Francisco to Los Angeles in 30 minutes. When the Hyperloop prototype was tested in Nevada all of the major news organizations in California covered the story because there was that much excitement over the idea that Mr. Goddard came up with more than a 100 years earlier. As a teenager Robert Goddard also invented the first gyroscope for airplanes while he was still in school. A gyroscope is a device that maintains stability commonly used for navigation and flight.

Dr. Robert Goddard was an American professor of physics and an inventor. He is credited with creating the first rocket fueled by liquid in 1926. He would go on to be the first scientist to establish that rockets in space cannot be powered by atmospheric oxygen because there is none space so the lack of it would eliminate combustion.

Between 1926 and 1940 he and his team launched 34 rockets, some traveled as fast as 550 miles an hour and soared to a height of more than a mile and a half, quite the achievement at the time. Mr. Goddard was also the first person to build a rocket that would travel faster than the speed of sound, with sound traveling at 768 miles an hour. He had the nickname of "Moony" possibly because of his ambitions to create a rocket that could fly to the moon or maybe because of his boyhood dreams when he would climb trees and imagine about traveling to the moon. Apparently the motivation from all of those science fiction novels would stay with him for life.

His work brought him into contact with the Germans who at the time were Nazis. Mr. Goddard tried to warn American officials of the impending threat but his warnings fell on deaf ears. Dr. Goddard was concerned in particular with work the Germans were doing on long range rockets. In fact, Robert Goddard unknowingly assisted German engineers who would call him on the phone about his work. Again, once Mr. Goddard realized the peril he cut off German communication and alerted the Americans, who at the time did not think the Germans could create a long range rocket capable of flying across the Atlantic and hitting American targets.

Dr. Robert H. Goddard died of throat cancer in 1945. He passed right at the dawning era of the jet and rocket ages and his significant accomplishments and insight earned him the title of one of the founding fathers of modern rocket science. I harken back to the talented people who refuse to accept the impossible as the unachievable and wonder what Robert Goddard would have to say about the concept of "sight speed"!

Neutron, Pulsar and Binary Stars!

It is thought that more than half of the stars in the solar system are not like the sun but instead exist in pairs which are called binary stars or a binary system. In this relationship two stars orbit each other, unlike the Sun which orbits the Milky Way galaxy by itself. While it takes the sun about 226 million years to complete an orbit, some of these binary stars complete their orbits in a matter of hours, albeit their orbital paths are much smaller, yet these stars complete their shorter orbits at astonishing paces.

In at least one case, four stars were found in very tight orbits. Sometimes the two stars that orbit in these tight formations close to form one larger star, when this happens it can result in a gamma ray burst. Gamma ray bursts are said to be some of the brightest blasts in all the universe and can be caused when two stars orbiting each other at super speeds are forced into one star before they explode.

These binary stars often form at their births and come from the same molecular cloud known as a nebula. We'll take a look at nebulae in a later chapter. As binary stars evolve their atmospheres often shrink as a result of the intense gravity caused by their super-fast orbits. Binary stars are important to science as they allow us to observe their influence upon one another as well as helping us categorize stars in general. Like normal stars, binary ones can orbit around black holes. They can also cause super nova's and some binary stars are invisible and only become visible when they exert their gravitational force on the other star. Binary stars are more common than originally thought and one theory is our sun is part of a binary system although the sun's partner has not been discovered, possibly because its orbit is too far away.

A binary pulsar star system was discovered in 1974 with one half of the pair being a very dense neutron star. Pulsar stands for pulsating radio star and is really just a neutron star that emits pulses of electromagnetic radiation. A neutron star is a very small and very dense star, thought to be what was left over from a super nova explosion. So these stars are all similar, again a pulsar star is a neutron star that pulses and both stars might have been left behind after a larger star exploded.

Since 1974 more than a half dozen binary pulsars have been discovered, which leads to something new called a bursting pulsar. Just a hand full of these bursting pulsars are known to exist and they emit some of the strongest x-rays and gamma rays on record. A group of high school students went on to confirm the extremely rare binary pulsar system. Many of these fast moving star systems have very tight orbits such as the one discovered by high school students. Its orbit was small enough to fill the space between Mercury and the Sun, remarkable because most of these stars have orbits that are even small and faster than that.

So how are these neutron and pulsar stars created? To the best of our knowledge they come about during a super nova and after the core of a massive star is compressed into a neutron star. The neutron star keeps most of his momentum and rotational speed and goes on to emit beams of light similar to a light house. If you have seen a light house at work its beams can only be seen when it is shining directly at you. The pulses from these stars come fast, some beam every four to five seconds, others beam more than a thousand times a second. Neutron could be the densest objects in the universe with one tea spoon of its material weighing one billion tons. Imagine packing the mass of Earth into an area the size of Chicago and you get a better picture of the general idea. Some are even more dense: try cramming two

Suns into Los Angeles, remember you can fit a million Earths into one Sun, so you could fit two million Earths into one Los Angeles and that would be the density of some of these stars. Scientists say if the material from a neutron star could be compressed into a cubic yard that yard would weigh more than the Atlantic Ocean.

So you can imagine something this dense is going to have a major impact on gravity. Anything that would fall toward a neutron star would probably be torn about by its tremendous gravity long before it made it to the surface. The gravity from the neutron star would accelerate the approaching object to almost a million miles an hour before it slammed into the surface. This is one of the many times when science is stranger that science fiction. Whatever was falling toward the surface of the neutron star would not be recognizable once it made it to the surface. This kind of gravity is so intense that it can warp light, meaning you might be able to see more than half of the star at any given point. It could be like the reflection from those mirrors that are curved which allow you to see around a corner.

Speed does make up for size, while neutron stars are relatively small they are super-fast. They spin so fast that the bulk of the star becomes compressed beyond comprehension. They can spin several times a second, or even faster. That would be like a planet the size of Los Angeles spinning so fast that it could complete three revolutions in one instant, or put another it completed three days in one second. And if another star comes too close, that stray star would lose its material to the spinning neutron star and in the process that neutron star would spin even faster. Now imagine that planet, the size of Los Angeles spinning 300 times a second. That could be equal to 300 days in one second, which is almost a year passing in a blink of an eye. They spin so fast that they actually start to bulge in the middle because of the centrifugal force. They do eventually begin to slow down but it takes a while. After the passing of a million years one of these spinning marvels might have only slowed by a few hundreds of a second.

Neutron stars could be more like planets than stars, with an extremely smooth solid crust due to its tremendous spinning speed. That surface is also extremely hot, around a million degrees once it has started to cool. The surface crust is extremely durable, maybe millions of times stronger than steel and flatter than a pancake because of that intense spinning speed and gravity. We're told there is no life on a neutron star, how could there be considering it is one of the most radical and unforgiving places in the universe.

The Mother of What!

Dr. Nancy Grace Roman is known as the mother of the Hubble Space Telescope, quite the prestigious tittle considering how much science has learned from the Hubble and all of the obstacles she had to overcome as a woman before the ambitious project was complete. Dr. Roman always knew she would be an astronomer. When she was four she drew pictures of the moon, her father was geophysicist and encouraged that sort of thing. Her accomplishments should not come as a surprise when you consider she was the only child of very loving, accomplished and encouraging parents, one a scientist and the other a teacher.

When she was 11 she formed an astronomy group at school, which was something unique for girls at the time. She continued to excel in school and years later she was offered a job at NASA, which was something unique for women at the time. One possible explanation for the NASA job offer was NASA needed to keep pace with the Russians who already had a woman in its ranks of government astronomers. NASA wanted Nancy Roman to build a platform for an orbiting telescope.

So like all good leaders and department heads, Dr. Roman consulted the best and brightest minds on the subject before making any decisions on her own, traveling the country to make sure she did not overlook anything or anyone. At that time astronomers were restless with Earth bound observatories, the consensus being looking at stars through the atmosphere was like looking through a tainted dirty lens with lots of defects. So she was really motivated to change this by somehow building a different type of observatory, one that might actually work in space.

She also became NASA'S first Chief of Astronomy. But her ideas always faced resistance and over the years she fought an uphill battle, facing skepticism from congress which was reluctant to fund her project. Some people say her biggest accomplishment was getting the money for the Hubble from congress, without which the Hubble may never have come to fruition. The consensus at the time was observatories on Earth were heavy and cumbersome and probably could not be launched into space. But after many problems and much time she designed what would eventually become the Hubble Space Telescope. We probably already know what Dr. Roman would have to say about "sight speed."

Dusty Roots!

The old saying "ashes to ashes and dust to dust" seems appropriate when applied to star dust from nebulae, which basically means we come from dust and we return to dust. A nebula is thought to be a large cloud in space were stars are formed and the cloud will eventually return to star dust just like everything else in the universe including us. More specifically, a nebula is a giant cloud in space made from dust, hydrogen and other particles. It is so big we really cannot imagine its size. They can stretch for hundreds of light years. Remember one light year is equal to six trillion miles. They can be formed when an interstellar medium collapses. An interstellar medium is the space between a solar system that is inside of a galaxy. When this medium collapses it is thought to compress part of the nebula and that causes the area to crumble which leaves behind these giant clouds of gas.

The can also be created when a really big star explodes which is called a super nova and scatters dust, gas and other particles into space. These particles condense and collapse under their own gravity and eventually create one larger part also known as a nebula. They are often called star nurseries because this is where stars are created. It is also believed that the first stars were formed from a nebula that was left over after the universe was created immediately following the Big Bang. So there are many theories regarding the formation of nebulae. For clarification, nebula is one cloud and nebulae is more than one or several clouds.

When those first stars died and went super nova they created a second generation nebula. The second generation nebula had heavier elements than the first so it created stars that were more complicated than the first generation stars. The more complicated the star, the greater the chance it will eventually form life. So after a few generations of these stars blowing up, returning to dust and blowing up again they are so heavy with enough of the right elements that simple life forms finally develop. The process took billions of years. The right elements include materials that are heavier than hydrogen and helium such as copper, oxygen and carbon to name a few. So the saying dust to dust seems to be a perfect description for this process.

Stars form inside a nebula in much the same way that the nebula forms from the dust cloud left behind from the Big Bang. Gas and dust inside the nebula are crushed together under tremendous gravity and they become denser and hotter until the hydrogen inside them starts to glow, becomes ignited and another star is born. It's also believed that planets are formed in much the same way. There are different kinds of nebulae and they can take some bizarre shapes, some are pictured to have protruding chimneys, while others might look like a cat's eye or even a lobster or a witch's head.

One type of nebula is called a planetary nebula but the name has nothing to do with a planet.

A planetary nebula is a ring shaped cloud formed by the expanding gas that surrounds an old star.

An example would be a star similar to our sun burns through its hydrogen and starts to burn its helium, that star will become much bigger right before it becomes a lot smaller which is called a white dwarf. The white dwarf has burned off its outer layer and only the core remains which still puts off lots of light and radiation.

In 1980 astronomers discovered a nebula that looked like a boomerang, so it was called the Boomerang Nebula. But when technology improved and the Hubble Space telescope was created, scientists took a second look at the Boomerang Nebula which now looked more like an hour glass. The boomerang apparently kept its original name but something else was learned. Astronomers also discovered it was in one of the coldest regions of space where the temperature was very close to 450 degrees below zero.

There is also something called dark nebula. Since about 90 percent of the entire universe is made up of dark matter and dark energy it only stands to reason that dark nebulae should be included. These interstellar clouds are so dense that they block out the light from stars that are in the background. They do this because their dust particles are coated with nitrogen that is frozen and along with carbon monoxide combine to block visible wavelengths of light. And just like the lighted ones, dark nebulae can also give birth to stars, planets and moons.

Remember that saying, "dust to dust", well apparently about five billion years from now our sun will return to its dusty roots and become part of a nebula, which will make future stars more fertile. Like most everything else in the universe, nothing is close, the nearest nebula is more than a thousand light years away. Obviously nebulae are denser than the space surrounding them but they might not be as dense as one might think. If a nebula cloud were the size of Earth, the nebula's mass might weigh a few pounds. So not only are they too big to comprehend but that are also probably too light fully understand as well.

There are many theories about the number of ways nebulae are created, many of the ideas posit nebulae consist mostly of hydrogen, helium, plasma and dust particles. Here are two of the most popular theories. A nebula is formed when an area of space is compressed and collapsed which leaves behind massive amounts of dust. The other theory is a nebula is formed by the explosion of a massive star.

Many of those beautiful pictures you see of nebulae in magazines can be misleading. Those pictures are sometimes altered so the entire nebula's mass can fit into one page or frame. This is extremely difficult to do when you consider those brightly colored clouds are so much larger than we could ever imagine. The bright spots in these clouds are the beginnings of new stars starting to form, still too hot for even the simplest forms of life to develop. Artistic scientists often paint parts of these pictures to illustrate the ultraviolet radiation that is emitted from these massive clouds.

One nebula in particular can be seen under the right conditions with the naked eye. It looks no bigger than the smallest spec in the night time sky near Orion's Belt but is more than 50 light years across, shinning down on us as a reminder that as we are in the universe the universe is in us while the cycle from dust to dust continues to play out.

Friend or Foe?

If we all get by with a little help from our friends then Jupiter is an extremely good friend, at least so far. It is said that the biggest planet in the solar system uses its considerable gravity to shield Earth from asteroids and other harmful material headed our way. For millions of years Jupiter has deflected asteroids and comets that otherwise would have smashed into Earth. And for millions of years Jupiter has also ejected asteroids and comets in our direction.

That's right, Jupiter not only throws asteroids and comets out of the solar system but it also hurls some of them inward toward the Sun where they have the chance of striking us on Earth. One example was in 1770 when one of these comets came within a million miles of Earth, a whisker hair by universal standards. A few years earlier that same comet soared by Jupiter which changed its orbit and sent it too close to our planet. A few years later the same comet flew back by Jupiter which this time hurled it in the other direction and out of the solar system. Seems Jupiter can be a fickle friend!

Another example came centuries later when the comet Shoemaker Levy smashed in to Jupiter. The collision was recorded and was used to convince congress to fund a program that would look for similar objects headed Earth's way. Had Shoemaker Levy impacted Earth, you might not be reading this book. There was another wakeup call when another object smashed into Jupiter with such force that the light from the explosion could be seen from Earth with a common telescope for a few seconds. It is not certain if that object was going to hit Earth but what is certain is, we should be glad it didn't. Some scientists take the question one step further by saying, if Jupiter was not there, life on Earth would not have had the chance to develop to the point where people have taken center stage.

The question of Jupiter is protector or menace could hinge on from where these large space rocks come. Some comets are long period while others are short period. The long period ones orbit the sun in elliptical patterns, where their paths take them out of the solar system and then back in every 200 years or so. The short period ones make their rounds every 20 years or so. Both types of comets go around the Sun and the likelihood of them hitting Earth could depend on how close they get to Jupiter.

Asteroids on the other hand pose a more consistent threat to Earth. The majority of these asteroids are contained in either the Kuiper Belt, near the outer solar system beyond Pluto, the Oort Cloud which wraps around the entire solar system or from the Asteroid Belt, located between Mars and Jupiter. Most of the asteroids stay in the Asteroid Belt without event. There Jupiter's gravity keeps them just active enough so they cannot form into a planet but in this state they can easily be disturbed by Jupiter's gravity and become thrust into the inner solar system toward Earth. So if the question of whether Jupiter is friend or foe is answered in the present, that would be yes, until of course the future has Jupiter hurling one of these objects into our happy home then the answer would be no.

It seems everything is bigger on Jupiter, including its ocean. Scientist believe that its ocean is the biggest in the solar system but it's not made of water. When Jupiter formed it retained higher levels of hydrogen and helium than other planets, in fact Jupiter could have been a star had it been 80 times larger. So all of its hydrogen and helium is compressed into liquid under the planet's tremendous gravity. The result is a liquid metal ocean so deep that it would completely submerge Earth. Or put another way Jupiter's ocean is as deep as Earth is round. This liquid metal ocean also creates the strongest magnetic field in the solar system, said to be some 14 to 20 times stronger than Earth's.

Jupiter's gravity is also more than twice that of Earth's, so a person weighing 100 pounds on Earth would weigh about 240 pounds on Jupiter. It also has 63 known moons, maybe more that we can't see. Its winds can blow 400 miles an hour, bitter and cold, with temperatures around 234 degrees below zero. You can see Jupiter with the naked eye and only the Sun, the Moon and Venus are brighter in the night sky. The first recorded sighting of Jupiter was around 8BC and more than a thousand years later we would learn that Jupiter has a faint system of rings that surrounds it.

So while Jupiter is known as the big Gas Giant, the fifth planet from the Sun and is more than twice as massive as all of the other planets in the solar system put together, we still don't know what to make of it. Despite sending a total of eight space craft to its surface we are not sure if we visited a friend or foe. It has the shortest day of all the planets in the solar system, meaning Jupiter turns on its axis about once every ten hours, we're uncertain if it will help or hurt us. Its Great Red Spot storm has raged longer than the United States has been in existence and while one of its 63 moons is larger than Mercury we can only deduce that Jupiter will continue to be our friend until it becomes our foe.

Much Bigger than Mt. Everest!

As big as Mt. Everest is it is not the biggest in the solar system, that title belongs to a mountain on Mars called Olympus Mons. Olympus Mons is so big that is seems to sort of distort the shape of the Red Planet, and it just might when you consider Mars is about half the size of Earth but its biggest mountain is about three times larger than Earth's biggest mountain. Imagine if Mt. Everest was three times bigger, it would reach nearly 20 miles high and be so big it might look deformed. That puts into context how big Olympus Mons is relative to the size of Mars. The mountain on Mars is so big that it would cover Arizona. It is 16 miles high and about 375 miles wide. Even by galactic standards that's a big mountain. It also has a volcano that could erupt again. Apparently volcanoes on Mars can be anywhere from ten to 100 times taller than those on Earth. The large volcano in Hawaii is called Mauna Loa and could actually fit inside Olympus Mons.

But Olympus Mons is not the kind of volcano that blows its top, instead when it erupts lava flows slowly from its sides which is called a shield volcano. For this reason, shield volcanoes are not as steep as one might think, their average incline is about 5 percent. At the top of Olympus Mons is an opening so big it would stretch half way across Long Island and cover some 53 miles. It is surrounded by cliffs that are six miles tall, which are as high as Mauna Loa at 13,678 feet.

So how did such a big mountain form on Mars but not on Earth? That answer, like so many other galactic questions may be found in gravity. Mars's gravity is about one third the strength of Earth's. So it is believed that Mar's lower surface gravity combined with its high rates of volcanic eruptions allowed the lava to create a larger mountain in the form of Olympus Mons. Another factor might involve tectonic plates. Earth has many of these underground plates which move on a regular basis and prevent lava build ups in a particular area. Mars has limited plate movement which allows for a greater accumulation of lava and bigger mountains. This led scientists to suggest if Mars had only one volcano instead four, that single volcano might be much larger than it already is and could be the highest surface protrusion in the entire solar system.

An incomplete and brief history of Mars is as follows. In the 1996 NASA reported that it had confirmed the existence of life on Mars. That claim was scrutinized by scientists all over the world and most concluded that life does not exist on that planet and there was not enough evidence to conclude that it ever did. Possibly not wanting to accept no for an answer humans have sent more than 40 missions to Mars with the Russians being the first to launch a probe in the early 1960's.

There are two functional robots currently on Mars. Both were sent by the United States which continued to search for signs of life that might be left behind in the sedimentary layer. They have discovered a huge sheet of ice almost one-foot-deep that stretches to about the size of Texas. The Lone Star State covers about 268,000 miles, so that's a big ice sheet. All that ice combined with earlier data from telescopes regarding what seems to be canals on the surface suggests that water did flow on Mars at some time in the past. But the bottom line is life as we know it cannot develop in the Martian atmosphere.

Mars is called the Red Planet because it is rusty, apparently the iron in its soil combines with oxygen and turns into iron oxide, or rust which gives it is red color. Mars also has two moons that are small by galactic standards, so small, about 15 miles in diameter that both of them could probably fit comfortably inside the county you call home. Mars has seasons just like Earth but the seasons on Mars are more extreme, but at risk of contradiction some scientists say during certain times Mars can be as warm as parts of Earth. Mars was visible to us long before the invention of telescopes, so humanity has had ample time to write and wonder about the planet which is most like our own.

The Biggest, Blackest of Holes!

Astronomers might have found the very first black hole and as you can imagine it is too big for us to really comprehend, might even be too big for the most brilliant astrophysicists to completely understand. This old black hole could be 12 billion times larger than the sun. It is called an "ancient monster" that could date back to the dawn of the universe or about billion years after the Big Bang. If it's not the first black hole we're told it is among the first group.

Scientists think these big black holes are at the center of every galaxy including the Milky Way. The black hole at the center of the Milky Way is thought to be half the size of the "ancient monster", remember the Milky Way black hole is almost five million times larger than the sun. The term black hole can be misleading, because some of them are bright. How can that be? Apparently black holes can be surrounded by something called an accretion disk or accretion disks. Accretion disks can be formed when an object like a planet or star gets too close to the black hole. These discs are made of dust particles and other matter that swirl around, downward into these black holes, gravity pulls them together causing friction and they heat up and give off light, illuminating the black holes so we may see them.

You can't really talk about black holes without talking about quasars. Scientists are still not quite sure what is a quasar even though the first one was discovered decades ago. Remember quasar is short for quasi-stellar radio source. Here's our current understanding of the phenomena. Quasars could actually be galaxies that are in a particular developmental stage and give off tremendous light and energy, thought to be among the brightest objects in the entire universe. Or put another way, quasars are thought to be extremely compact areas situated in the middle of a super large galaxy and at the center of these light producing quasars are black holes. That's about the most direct, simplest definition I can find. Again they are so complicated that even the smartest of scientists don't fully understand them.

The current number of known quasars is disputed but scientists are certain that none can be seen with the naked eye. It is also thought that each quasar has its own black hole and most of these black holes are about 1 billion times the size of our sun. The exception is that "ancient monster" recently discovered which again is about 12 billion times the size of the sun. It is almost 13 billion light years away. Remember this light travels from the past, so it is about 13 billion years old, the universe is about 13.8 billion years old so for this reason the "ancient monster" could be the first one created.

All this took place when the universe was just some 900 million years old, still in its infancy. In theory, that would be just enough time for the Big Bang to go off, which created the first stars, which would later collapse and eventually create the first black hole. It only stands to reason that with the biggest black hole in that "ancient monster" you would also have the biggest quasar and that's what scientist think they have discovered. Remember quasars are said to be some of the brightest objects in the universe and this quasar is thought to be more than 400 trillion times brighter than the sun. Can humans comprehend anything brighter than the sun? The sun's so bright we can't even look at it without risking blindness.

Despite the significant knowledge accumulated about black holes, scientist say they still don't have an adequate theory to explain how they formed when the universe was in its beginning stages. But they do know enough to say with confidence that black holes are formed when stars that are much bigger than our sun collapse. It is estimated that the stars that create black holes must be at least 40 times the size of our sun.

So is Earth in danger of being consumed by the black hole at the center of our galaxy? According to scientists, no. Apparently earth is not close enough. As that black hole sits in the center of the Milky Way, Earth is situated about two thirds of the way out, again too far away to be in danger.

Black holes can also effect time which is an appropriate place to introduce Albert Einstein's Theory of Relativity. While it is much too complicated to detail here, it suggests that time and space are related to the energy and momentum of whatever matter or object is present in the area. Black holes warp light. Black holes also warp time. Think of a heavy person sitting on a mattress. That mattress is warped by the person's weight. Consider the impossible scenario of two people near a black hole, one was a bystander and the other a victim. As the victim was pulled into the black hole his momentum would appear to slow down and the hands on his wrist watch would also appear to slow down. The victim would not notice these changes but the by stander would. This is not quite science fiction but it is one of the more radical theories of Albert Einstein.

As our knowledge of black holes increases we risk getting slapped with reality, similar to the slap we got when we first traveled to the moon. During those early lunar missions humanity was so ambitious, we entertained dreams of traveling to distance planets and even stars. But when the reality of space travel became apparent, we realized that we just couldn't traverse those great distances throughout the solar system, let alone the galaxy or universe. This limitation did not stop our efforts to travel in space, it just sent us into new directions. For example, we became more dependent on unmanned probes that were sent were humans could not go. We used space telescopes to learn from a great distance what we could not learn up close from direct observation. If this is the fate and that awaits us when we learn more for black holes?

The Universal Icon!

Two of the most iconic symbols of space are said to be Saturn and the Crescent Moon and in that order, so we will examine the former in some detail. Saturn is huge, the second largest planet in the solar system right behind its giant neighbor Jupiter. Almost ten Earths would fit across Saturn but don't be fooled by Saturn's great mass because it could be called a light weight. Saturn is considered to be the least dense of any planet in the solar system, an amazing planet unlike our own and more similar to the Sun than Earth in its composition. Saturn and the Sun are mainly comprised of hydrogen and helium. Earth contains many more elements such as rock, iron, magnesium, calcium, carbon, oxygen, and too many more to list here. Saturn is one giant ball of gas, it has no solid surface on which you could stand and it is lighter than water. Scientists often joke that Saturn could float in your bathtub, if you had a large enough tub.

Those beautiful rings of Saturn are also not solid but instead are made of dust, rock and ice. Some of these particles are as small as a grain of sand, others are more than one half mile wide. In relation to its body, Saturn's rings are razor thin. If Saturn were the size of a baseball its rings would be would be much thinner than the thickness of a strand of human hair, too thin to see. If you could put the rings in a line they would almost stretch from Earth to the Moon, a distance of 238,000 miles but those rings are less than a one half mile thick, making them razor thin in relation to their length. Saturn has seven rings and they could all disappear in 50 million years or so. Scientists say the rings will either be pulled into Saturn because of the planet's significant gravity or they could just dissolve into space.

Saturn is titled on its axis so when it is at the right angle its rings cannot be seen from Earth. That's because the rings are seen straight on and seem to disappear, again in part because they are so thin. Saturn reaches this angle about once every 14 years and it caused problems for at least one early astronomer who first saw the rings and then saw them disappear only to reappear years later. We're told that early astronomer thought he might be losing his sanity. Saturn also has long summers, about eighty years long apparently caused by its tilt. It is a gassy, windy place that would offer no quarter for humans. Winds on its surface can exceed 1,000 miles an hour. In those conditions you would not be blown away, you would be ripped apart. Think about what exposure to the outside of a jet liner at top speed would do, now double that and you a better idea of Saturn's destructive wind force. Those high winds can fuel storms that last for months or years and apparently some of those storms are still raging on Saturn's surface.

Scientists recorded one long lasting storm on Saturn that produced lightning purported to be more than 1,000 times stronger than on Earth. A few years earlier scientists reported an ammonia storm on the planet which was about five times bigger than something called "Snowmageddon" on Earth. Remember "Snowmageddon" it was that storm that closed Washington D.C. in 2010.

While Saturn goes around the Sun slowly, it spins fast. It takes the planet 29 years to orbit the sun but it spins on its axis once every 10 hours and fifteen minutes. Or put another way Saturn's year lasts 29 years but its day lasts 10 hours. And because it spins so fast and is made of gas, Saturn is flat on the top while its center is bulged out. Saturn is a long way away at an average distance of some 886 million miles out, it is about eight times farther from us than is the Sun. Remember the sun is about 93 million miles away. Put another way if you were to drive to Saturn in a car at 70 miles an hour it would take about 1,300 years to reach the gas planet which would give you a cool reception with temperatures of about 275 below zero. Saturn is so distant that if we viewed the Sun from its surface the Sun would appear ten times smaller than it does on Earth. It is so far away and so unlike Earth that it has been visited only four times by robots and other probes. Three of those four visits were just short fly byes. But one of those fly byes was long enough to confirm that Saturn has some 62 moons and that figure could go up. In fact, one of Saturn's moons holds more promise for life than does Saturn. That moon would be Titan. Titan is the second largest moon in the solar system and is home to a large body of liquid methane.

For years scientist thought the secret to life was liquid water. Now scientists think one of the keys to life may be just plain liquid, either hydrogen, oxygen or even methane to name a few. Its seems there is a certain quality about liquid that allows small elements like cells and molecules to bond together in a way that they cannot do on a dry surface. Titan is the only known moon that has a significant atmosphere, believed to be about 370 miles deep and is about 10 times thicker than Earth's atmosphere. Earth's atmosphere is less than 60 miles deep.

Saturn has another large moon that holds potential for life called Enceladus. Enceladus has an icy otter shell with water underneath and beneath the water is a rocky core. Some scientists say Enceladus could hold the most promise for holding alien life in the solar system. Problem is, Saturn and all of its moons are in a very cold place, temperatures are hundreds of degrees below zero. Despite this deep freeze scientist say there must be some sort of unseen warming process that prevents water on Enceladus from turning to ice. So while that strange heating mechanism remains a mystery, scientist know that Saturn gives off about twice as much heat as it receives from the Sun. One theory is the helium that sinks into Saturn's core can create a process that heats Saturn's core to maybe 20 degrees Fahrenheit.

Atmospheric pressure deep in Saturn's body is tremendous, about 100 times greater than Earth's. Imagine pressure so strong that it compresses gas into liquid, in this case hydrogen into liquid metallic hydrogen which of course is strong enough to crush almost anything man-made. We hope to know more Saturn in the coming years because another visit is planned to launch in 2020. That project is called the Titan Saturn System Mission or TSSM and is expected to cost 2.5 billion dollars.

The Titan Saturn System Mission!

Coming to our solar system in the near future could be the most ambitious plan yet to visit a distance planet and its moons. It is called Titan Saturn System Mission, created in 2009 with a planned for launch in 2020. It will cost at least 2.5 billion dollars and it's a joint effort between NASA and the European Space Agency or ESA. Saturn is the second largest planet in the solar system and Titan is one of Saturn's many moons.

The TSSM is scheduled to arrive at Saturn's moon system in 2029. It's a nine-year trip, that will spend four years at is destination. Once there, half of that time would be spent touring Saturn and the other half studying its moon system. When Earth and Saturn are nearest each other the trip to Saturn is 746 million miles. It will require four gravity assists. Remember gravity assists work just like a sling shot between a planet and a space craft and are used to increase speed or change direction of that craft. The craft will enter Earth's gravity and speed up significantly and then exit the gravitational field with the "sling shot effect" of having much more speed than before. The same procedure will be done with Venus. Remember Earth is traveling around the Sun at 67 thousand miles an hour and Venus is traveling even faster at 78 thousand miles an hour. So these assists are necessary for the craft to reach Saturn on time.

The mission will carry an orbiter and two probes to explore Titan, one of Saturn's 62 confirmed moons. We often think of moons being smaller than planets but Titan is larger than both Mercury and Pluto. The mission will also carry a hot air balloon that will float in the clouds above Titan. There will also be a lander designed to splash down in a sea of methane and liquid hydrocarbons. It is a sea that seems to be completely different than the ones on Earth. Hydrocarbons can be gas or liquid and can be comprised of hexane, benzene, polyethylene and the likes, these are chemicals that under certain circumstances can be fatal, unlike Earth's ocean which is life giving.

The lander is designed to be released from the orbiter during the second fly bye and will target the sea near the northern pole of Saturn's moon. Once it is released from the orbiter the probe will drop down through thick clouds in a descent that will last six hours. The six-hour trip should be plenty of time for the probe to collect and analyze data from the cloudy atmosphere. With its long lasting batteries and a specially coated outer shell the probe is expected to last up to six months floating on the surface of a methane sea and collecting data that will be sent back to Earth.

When it gets to Saturn the probes and robots are expected to perform a number of fly byes of moons Titan and Enceladus. Saturn and its moons are too far away from the Sun to use solar power so it will use special batteries known for their abilities to store energy for long periods. The probes and robots will take a closer look at the moon Enceladus which is said to be more reflective than snow. Here is the explanation, Enceladus is covered with fresh, clean pure ice which reflects most of the sunlight that hits the surface back into space. So the temperature during the warmest part of the day on Enceladus would only make it to about 200 degrees below zero.

The mission promises to be record breaking and is part of what's called the first floating exploration of an extraterrestrial sea. Put more directly, it will be the first mission to explore an ocean in space. Scientists are curious to see if there are any similarities between Earth's oceans which are made of water and chemicals like sodium and those on Saturn and its moons which are made of methane and chemicals like polyethylene. Scientist would also like to know if

B.E Chenault

Saturn's atmospheric and oceanographic systems are intertwined in a way that would be similar Earth's natural system. Titan is said to be very Earth like and might hold some of the best potential for finding alien life in the solar system. Once again those answers might come from a trip that is literally out of this world with the potential coming from a moon instead of its planet.

A Worm What!

A worm hole is a theory and not reality but the concept comes from the mind of Albert Einstein so it has gathered significant attention down through the years. It seems that the radical concept of a worm hole is not that much different from the concept of "sight speed" in that both try and reconcile the great distances of space with the short life spans of humans. But you would think that such a sophisticated concept would deserve a more respectable name than worm hole.

A worm hole is a theoretical concept where a passage in space creates a short cut in order to traverse the tremendous distances of the universe. Think of space as one flat piece of paper and at each end of the paper are two dots. The distance between those two dots is too great for human travel but if you fold the paper in half then those two dots are right on top on each other. That's the simplest way scientists can describe the concept of a worm hole. The theory posits that the universe might be manipulated like the sheet paper.

Again it is just a theory and it could not be used to human travel. According to Mr. Einstein as soon as a person or object entered the opening of a worm hole, part of the worm hole would collapse and kill the would be time traveler. Scientists can't explain why part of the worm hole would collapse. Worm holes are also small, too small to transport people. According to the theory, a worm hole that naturally occurs in space would be more than a million times smaller than a single atom, far too so small to even send signals. Remember there are several billion, billion atoms in the human hand.

It is thought that worm holes might be able to access parallel universes or different time periods within a parallel universe but because they are so unstable they would instantly collapse if any material, no matter how small attempted to pass through it. A worm hole is thought to be a beam of light that passes between two points in time that has a curved space. Again they would not be useful for travel because they are too small and would break apart when something tried to enter.

The creation of a worm hole would require the control and redirection of large quantities of matter. Some scientists theorize that a worm hole just one yard in mass would require the manipulation of a huge planet more than a thousand times larger than Earth. One can only imagine the problems of trying to squeeze a planet a thousand times larger than Earth into a yard. You can also imagine we're not even close to being able to do this, which probably why it's called a radical theory as opposed to just a regular theory.

There are more problems with this theory, including the intense gravity caused by such an attempt.

Trying to squeeze a large planet into a small area would require manipulating something called exotic matter. Exotic matter is matter is unlike regular matter. Regular matter contains protons and neutrons and exotic matter does not. Fact of the matter is scientists don't know what makes up exotic matter, they just think it is matter that somehow is different from regular mater because it has exotic properties. There is no evidence that this matter exists just like there's none to suggest worm holes exist.

So for the present, worm holes only exist in science fiction and in the imaginations of brilliant scientists who continue to think outside of the box. But these brilliant minds are not deterred by the great challenges and they continue to investigate the possibilities of the impossible, such as worm holes and space travel. Had someone told the scientists of the previous generation that humans cannot fly 100 miles above Earth at speeds more than 15 thousand miles an hour, we might not have ever made it to the Moon. So while these are just theories and some are radical theories who knows where they can lead and what benefits they might produce.

That's a Long Time!

A Russian cosmonaut holds the record for the most time spent in space. His name is Gennady Padalka and throughout the course of his distinguished career he managed to log an amazing 879 days in space. That's the equivalent of about two and a half years in space, zero gravity, motion sickness, bone and muscle loss and everything else that comes with it. Although those days were not consecutive it is still an amazing accomplishment. He began his career in 1989 and broke the record for the most time spent in space by any human in 2015. Gennady Padalka would like to eventually spend a 1000 days in space.

Science tells us that the human body begins to immediately break down in space, quickly losing bone density and suffering muscle mass. There are many short term effects like damage to the eyes and the long term ones have yet to be fully understood. Living in space is just not natural, so you can imagine doctors will be monitoring Mr. Padalka closely, not only for the benefit of his health but also for the advancement of science. Since the very beginning, humans and everything else living on the planet have all adapted to and depended on gravity, including even the smallest of microbes. So it is sufficient to say we still don't fully understand how humans will be affected by the long term effects of living in space.

Upon returning to Earth some astronauts experience health problems which were expected. One man was not permitted to drive for weeks after he spent several consecutive months in space. And for a while he could not stand in the shower when he was back on Earth. He also had to wear a gravity suit to keep blood in his head, after spending so much time in space the body produced less blood to be circulated though out the body. There were long term effects on load bearing bones that lasted for years which also made those bones easier to break or crack.

Because astronauts tend to float instead of walk in space their load bearing bones in the lower part of the body tend to suffer the greatest atrophy. These bones break down and release calcium into the body which is reabsorbed. This condition leaves those bones more brittle and prone to breakage. This can result in the loss of up to two percent of bone mass in the affected areas of the body, which again are usually the lower extremities. The same process also has a negative effect on the muscles in the lower body. Weightlessness can also impact bodily fluids. In space blood and other bodily fluids tend to collect around the head because there is no gravity to pull them to the lower extremities. This can result in a feeling of constant congestion. And because the heart has a tenancy not to work as hard as usual in space the heart can actually decrease in size and as a consequence become weaker. The human heart is both a muscle and an organ and with regular exercise it actually becomes larger and stronger and without regular exercise just the opposite can happen. A weaker heart can lead to or accelerate any number of cardiovascular problems.

But astronauts do tend to grow in space, caused by the lack of gravity. With no gravity constantly pulling down on our bodies our spines expand and heights increase. On Earth the spine and its vertebra disks are compressed by gravity. In space there is no spinal compression which causes it to lengthen and the astronaut's increase their physical statures but the condition is not permanent and they return to their normal sizes after returning to Earth.

Space can also impact the equilibrium of the human body, much of our orientation and balance comes from senses in the ear. But in the weightlessness of space the ear does not function as it was intended which can cause problems for astronauts in their balance, their ability to stop and turn and other behaviors associated with general orientation, all of which are taken for granted on Earth.

To combat some of the negative effects of weightlessness, astronauts traditionally exercise in space for a couple of hours every day. Some astronauts also wear heavy clothing to stimulate muscle growth. But bone loss, muscle atrophy and suppressed immune system responses are problems that are not going away regardless of how much exercise and heavy clothing are factored in to the equation.

Of course astronauts outside of Earth's protective magnetic shield are more susceptible to cosmic radiation, the shield blocks out about 99 percent of the harmful radiation from the Sun. Exposure to that radiation increases the chances of cancer while decreasing the way the immune system functions. Often times the effects of radiation are low and can be mitigated by fortified suits and special compartments inside space crafts. But the risks of these dangers goes up in the presence of solar flares. Solar flares are explosions on the sun's surface which dramatically increase radiation levels. Because solar flares often give some sort of advanced notice or waring, humans in space might have enough time to lock themselves inside fortified compartments in their space craft. These special rooms might contain large amounts of water. Turns out water is an excellent insulator against cosmic radiation, possibly because of the large amounts of hydrogen in water. Ironically the sun which emits the deadly cosmic radiation is also made up largely of hydrogen so it seems in this context hydrogen can be both beneficial and detrimental.

Can you imagine spending more than two years without ripping or tearing your jeans, shirt or jacket on something like a car door, or desk drawer! Well imagine if you ripped your space suit which could be a fatal mistake in space. An astronaut's space suit is like a miniature ship, protecting them from the freezing conditions one moment and then inferno like temperatures the next. Besides regulating body temperature, the suit also provides the fundamental function of allowing the astronaut to breath. If there was a rip or tear in the suit while the astronaut was in space, there's the backup system of an oxygen tank but that would most likely only prolong the inevitable for a brief period. Space suits also provide protection from harmful radiation that comes from the sun and other parts of the universe. The radiation can come in the form of subatomic particles that are moving at super high speeds and can tear through our DNA molecules or split them, which again can cause cancer and other diseases. So two and years is a long time not to have any kind of significant mishap.

The more time you spend in Earth's orbit the greater the chance that you will be inadvertently hit by what's called space junk. Space junk is defined as almost anything in space, from debris left behind from rockets, satellites or fragments of space material and all of the space junk was deposited by humans. More than two thousand satellites have been launched and many of them or their parts are still in orbit just waiting to collide with something else. Humans began launching rockets into space in 1957 and since then it is believed there could be more than a million bits and pieces of those missions still up there in orbit blasting by at some 17,500 thousand miles an hour which pose a threat to any astronaut who is unfortunate to be in the right area at the wrong time.

And there is the natural space junk. We know about asteroids but did you know about micro asteroids, they're smaller versions of asteroids and bombard our atmosphere every minute of every hour. If you were ever hit by a rock that was thrown when you were a child, then you probably still remember the pain. Now imagine the same sized rock traveling around 20,000 miles an hour when it hits an astronaut working on the outside of the International Space Station. Chances are it would go right through the astronaut leaving the victim with a short time to remember that moment in his or her childhood when they were hit with a similar rock so many years ago.

Living in the Sun!

In one technical sense you could say we live inside the sun, if you consider something called the heliosphere. That's the term for the sun's outer atmosphere which extends much farther that what we can see. The heliosphere is also called the Sun's outer atmosphere and is defined as the entire region of space that is influenced by the Sun and its magnetic field which of course includes Earth and points beyond.

Some scientists define the heliosphere as a bubble around the entire solar system powered by the sun's winds which are called solar winds and can blow several million miles an hour. Depending on what planets, moons, asteroids or other space bodies are in the way, solar winds can travel great distances. If these winds are not blocked by any objects they can travel to interstellar space. This area of space could be one of the bleakest areas in the universe, so far out that it is not occupied by solarsystems, stars nor planets. It's the area in between galaxies and scientists tell us there is really nothing there, hence the name space. If a cubic foot of Earth's atmosphere contained one billion atoms, interstellar space might contain only one. Again it is not certain how far the solar winds travel but it is believed that they blow well past in the boundary line of our solar system which is said to be 1000 times farther away than Pluto. Pluto is more than 4.5 billion miles from Earth.

The heliosphere is thought to extend maybe nine billion miles. Scientists say proof of that might be the Aurora Borealis lights that have been see far out in the solar system. Because of the sun's motion in space the heliosphere is thought to be shaped like a tear drop with a tail that follows for billions of miles. Plasma that is ejected from the sun is shot out into the solar system and is thought to maintain the outer perimeter of that tear dropped shaped bubble called the heliosphere. When these winds finally slowdown in an abrupt fashion it is called termination shock. Termination shock is similar to when water from the tap hits the sink and starts to spread out, or when the brakes are applied to a car and everyone inside is pushed forward in their seats from the rapid deceleration. The short definition of termination shock is the area where the sun no longer has any influence.

The outer edge of the heliosphere could be called the outer edge of the solar system, so far out that you would not feel the sun's heat. Scientists think the heliosphere might move through the galaxy like a boat, with the bow and stern made up by the tear shaped heliosphere. They also believe that this bubble shaped heliosphere protects the solar system from cosmic rays and other harmful particles. There is also concern that this protective shield is weakening. Astronomers noticed the sun's solar winds have slowed to all-time lows, with the solar winds slowing there's concern the strength of the heliosphere will also weaken which might expose humans to greater amounts of radioactive material that can cause skin diseases in general and cancer in particular. The heliosphere seems to fit into the overall concept that everything in the universe is connected, the heliosphere surrounding the solar system, the solar system surrounded by the galaxy and the galaxy surrounded by the universe and all of it is compromised of the same basic matter.

What's in a Name?

Uranus's name is one of many strange things about that planet. Apparently there are now two pronunciations of the planet's name after the original pronunciation was known to be the source of embarrassment. One pronunciation sounds just like you think, "your anus" and the other "urine ess" is more discrete. Those who lecture young audiences are urged to use the latter pronunciation.

It was one of the last planets to be discovered in the 1700s and for some unknown reason it was almost named Neptune. At some 2.8 billion miles from the Sun, Neptune receives about one 400th of the energy that Earth gets from the Sun. So it's cold that far out, about 350 degrees below zero cold, thought to be the coldest planet in the solar system. But don't be deterred by the cold, it's also windy, gales blow more than 500 miles an hour. Back to Uranus, which is the fourth most massive planet in the solar system, some four times the size of Earth and it is the farthest planet from our home that we can see with the naked eye.

Another unique feature of Uranus is its position, it seems to be rotating on its side after it was turned over, possibly by some sort of cosmic collision. Because of its strange tilt, some people say Uranus is rolling around the Sun on its side. It was also the first planet found with the use of a telescope. The theory of the tilt involves a planet about the size of Earth that collided with Uranus several billion years ago. The impact was not direct because that would have likely destroyed Uranus, instead the collision was more of a glancing blow that left the planet in its current, sideways position. The impact theory may also hold the answer as to why Uranus is so cold. Scientists theorize that the Earth sized planet that collided with Uranus knocked much of its interior out into space and it was that interior that helped to generate the planet's heat. It's common for planets to generate heat from their cores. One of the more commonly accepted theories is that Uranus was struck by two smaller objects as opposed to one larger one. Scientists say the theory containing two smaller objects would better explain the positions and behaviors of the moons orbiting Uranus.

Next to Saturn, Uranus is the least dense planet in the solar system. If you could put Saturn in a bathtub it would float. Uranus is not quite that light but is very close and might bob for some time. While Uranus is more than four times as massive as the Earth, its gravity is much the same. Of course you cannot stand on Uranus because its gas, but if you could, your weight would be close to what it is on Earth, so if you weighed a 100 pounds here you might weigh 86 pounds there.

Uranus's more than 90-degree tilt makes for some long seasons, like 20 years long. No other planet in the solar system is titled more than 30 degrees. So while all of the other planets spin like tops, Uranus spins on its side. Because it spins on its side, a single day on its north pole lasts 84 years but its day at the equator lasts just 17 hours. It is one planet where one day has two different lengths, one lasting almost a century and the other not even 24 hours. Again fact can be stranger than fiction with a planet where one day lasts not even a day but at the same time that same day lasts almost a century. So if you were on its north pole you could see the sun rise, circle around for 42 years before it sat below the horizon after which there would be another 42 years of darkness.

At last count Uranus has 27 moons but that figure could go up as technology improves. Case in point, the last two moons of Uranus were discovered in 2011 and 2012 by the Hubble Space Telescope.

Even though it has too many moons to mention in this brief work, Miranda is said to be its most fascinating moon. The surface of Miranda is varied and full of craters, terraces and ice canyons. It is also the smallest and the closest of the moons that orbits Uranus. It was also featured in a Shakespearean play and a Hollywood movie, not to mention the chemical element Uranium which is used as fuel for nuclear reactors was named after Uranus. Next to Saturn, Uranus has the second set of most beautiful rings in the solar system.

Besides the strange name, the strange length of days and some strange weather, Uranus also rotates in what's called a retrograde direction which is opposite from the way Earth and most of the other planets spin on their axises. It also has rings that are so faint they were only recently discovered. They were found in 1977 and are made up of dust, rock and other particles. Uranus has been visited just once by the probe Voyager II in 1986. So the planned TSSM visit in 2029 will be something of a solar system milestone more than 40 years in the making. A little later we'll examine another humorous but still scientific side of Uranus and its possible odor that would provide material for late night comics.

One Big Anonymous Hole!

As the universe continues to expand what was the biggest or fastest will be relegated to a secondary position so it is predictable that the largest black hole which was recently discovered will probably not have that status for any length of time. This big black hole does not even have a name, it's just referred to by its host galaxy, called NGC 1277. Some of these figures are just too large to comprehend including this black hole, said to have 17 billion times more mass than the sun. Remember about one million Earths could fit inside our Sun so it really is hard to comprehend the size of this black hole. It is also the heaviest black hole ever measured and if it were at the center of our solar system this black hole would take up everything, including the empty space beyond.

Scientists are not sure what NGC 1277 might look like but think it could resemble a bowing glass lenses with one side in the middle that is thicker but tapers away at the edges. We will take a closer look at this a later but the assumed shape could challenge some fundamental believes about the universe that were established by Albert Einstein.

The giant black hole was discovered in the fall of 2012 and could have already been superseded as the biggest of its kind. It is about 250 million light years from Earth and takes up most of the mass of its host galaxy. Normal black holes take up less than one percent of that mass. It is so big that it is about 11 times as wide as Neptune's orbit around the Sun, so large that it might spawn a new class of black holes. Researchers were so stunned by the size of the black hole that they took an extra year to double check all of their measurements. Astronomers once thought that the size of a galaxy and the black hole at its center were related to each other. That conventional wisdom is now being called into question solely based on the size of this big unnamed black hole. This anonymous black hole takes up about 59 percent of the mass of its host galaxy. The next largest black hole, which is much bigger than the common ones, takes up only about 11 percent of its host galaxy's mass. It is clearly the black hole with the largest mass to galaxy ratio which is about 100 times greater than average ratios of this kind.

The big black hole is also in an area known for big objects, several other galaxies in the vicinity have the same sizes and compositions which could lead to the discovery of more of these giant wonders. No worries about them destroying our galaxy because they're too far out, about 250 million light years away. Finding these behemoths was not easy. First scientists studied light that came from 700 galaxies. Then they found about a half a dozen smaller galaxies with stars inside them that were moving at super-fast speeds. The high speeds inside of small galaxies were an indication that giant black holes might be present. Plus, these areas seem to be home to only old stars that are twice the age of our sun.

According to one theory, these old black holes were created shortly after the Big Bang and since then they have not undergone much change. If this is the case scientists are hopeful that these old black holes can tell us a lot about the early universe. Another theory is that the monster black hole was created from the collision of two smaller galaxies instead of one larger one and each of the smaller galaxies had a black hole of their own which formed into the present monster. Researchers say three things probably needed to happen in order to form such a large black hole. The occurrence of just one of the requirements is rare let alone three happening in a required sequence. First the black holes have to merge, then they must get ejected from one galaxy and last they must get caught by another galaxy. These are just theories but either way scientists are marveling at what they call the biggest black hole of its kind ever found, until a larger one is discovered.

There's a White One!

After hearing about the mysterious and powerful nature of black holes it may appropriate that we hear about their opposites, white holes. It is thought that black holes consume matter with such gravitational force that nothing can escape. White holes do just the opposite and repel matter with such force that nothing can get in. As we know black holes do their work with the help of gravity. White holes do the opposite and explode instead of imploding.

The theory of a white hole could be every bit as intriguing as the one about a black hole. According to the proposition white holes also have gravity that would attract you to them but the closer you got to one the harder it would be to move forward because of all the matter and energy being expelled. In theory you would never reach the white hole because of its tremendous outward thrust. This hypothetical approach has been compared to air resistance on Earth that slows down moving objects. The more air resistance means the more energy that is required to move forward, which can be translated to the closer one gets to the white hole the harder it would be to proceed. Scientists say nothing can reach and enter a white hole.

We are also told that space and time are warped around a white holes and approaching them has also been compared to going uphill. The higher you climb, the more energy is required until you reach a point where all of your energy is expended and you're at a complete stand still. Theorists say there would not be enough energy in the universe to allow you to enter a white hole.

So where do white holes get all this energy? Scientists don't, which makes them skeptical that they even exist. But according the theory of general relativity white holes should exist in some shape or form. The theory has two branches, the speed of light is constant for all people or objects and the laws of physics are the same for all people or objects that are moving at relative speeds.

Black holes are thought to form from something called gravitational singularity. It is a single point in space where gravity is thought to be so strong that it is infinite and really cannot be measured other than to say it is really strong and its tremendous strength attracts anything that gets too close. White holes work under the same general premise of gravitational singularity but in a reverse manner. Using this as a model, scientists theorize that anything that gets too close to a white hole will be blasted away. The same theory has been applied to gamma ray bursts, pulsars and even old black holes that have run their courses, anything that gets too close will be blown into oblivion.

No white holes have ever been observed, which puts further doubt on their existence. There is the postulation that the white hole is so big that we cannot see it, similar to being too close to the tree to see the forest. If white holes do exist than they would probably decrease entropy. Entropy is the measure of order in a system, or put another way, it describes the amount of matter in a system that is capable of movement that puts off heat. All living and mechanical things put off some measure of heat. So if white holes decrease entropy that would mean that white holes are slowly working to bring about the end of the universe. This sounds similar to the theory of the heat death of the universe which will be discussed later.

Taking a closer look at white holes raises even more questions. The theory of general relativity suggests that white holes could hypothetically exists somewhere in space and that they cannot be entered from the outside, as mentioned earlier because of all the matter and energy they eject. This theory suggests there might be some sort of way to enter a white hole. Consider all of the matter inside a white hole that gets ejected outward, that material must have gotten inside somehow. Remember a black hole has a "one way" in only entrance which is its event horizon which can be compared to a door. So why would not a white hole have something similar but only in reverse, maybe something like osmosis where sub atomic material can sort of leech in over time?

Would white holes violate the first law of thermodynamics? The law posits that energy can be changed from one form to another but it cannot be destroyed nor created and the total amount of energy in the universe remains constant regardless of its changing form. Remember white holes are said to eject matter but we don't know from where this matter came or what becomes of that matter after it is ejected.

Despite the lack of evidence that would support the existence of white holes there are some unexplainable phenomena that add to the mystery. In 2006 there was a gamma ray burst. Remember gamma ray bursts are explosions that create tremendous amounts of light, heat and energy. This particular gamma ray burst did not fit the current model to explain its origin. These bursts usually don't last long and neither did this one, lasting only for about 102 seconds. According to conventional wisdom it probably came from a super nova but there was no super nova in the area. After considerable research, observation and discussion, the question was raised, did this gamma ray blast come from a white hole? Researchers said they didn't know but were certain they had just advanced upon new and uncharted territory. That considerable research, observation and discussion continued for years until scientists were hit with an epiphany, did they witness behavior from a white hole? The tremendous force of that explosion that lasted less than two minutes might fit nicely into a white hole theory. The white hole ejects massive amounts of matter, heat, light and energy and then a short while later collapses on itself which causes the brilliant and explosive light show. Still there were more questions than answers with a large section of the scientific community stating something like this borders on the impossible.

Would not Newton's third law of motion apply here? It suggests that when a body exerts a force on a second body the second body simultaneously exerts a force equal in a size and opposite in direction on the first body. Even though the white hole theory is incomplete at best, it is still under consideration until a stronger one comes along.

The Biggest and Brightest!

Hypergiants are stars that could also be too big for us to comprehend. Some of these massive stars would take up most of our solar system. Imagine a star so much bigger than our sun that the big star takes up everything in the sky and there is nothing left. One such star is VY Canis Majoris and consider ourselves fortunate that VY Canis Majoris is not close because we would not stand a chance with it in the vicinity. It is so big that the star's mass would swallow up most of the planets in the solar system with its mass extending well beyond the orbit of Saturn. That's about 900 million miles, so of course Mercury, Venus, Earth, Mars, Jupiter, Saturn and beyond would all be well inside this massive star, most likely seared beyond recognition in its super-heated plasma. VY Canis Majoris is more than 1,500 times the size of the sun. The sun's circumference is measured in millions of miles, VY Canis Majoris's circumference is measured in billions of miles. It would take more than a thousand years to fly around it in a modern jet liner, the same trip around the Sun would take seven months and the same trip around Earth would take about a day and a half. VY Canis Majoris has big violent storms that would destroy most the material in the solar system. The Sun's storms pale in comparison which only disrupt electrical and communication systems on Earth. These hypergiants are also known for brightness or luminosity. The bigger the star the brighter it usually shines. Some are so bright and generate so much radiation that parts of its outer layers might get thrown off during massive explosions called coronal mass ejections.

One of the most massive stars ever recorded is called Eta Carinae. It could be 130 times more massive than the sun and it shines six million times brighter. Occasionally this star would exceed certain limits and throw off some of its outer core in massive explosions. One reason why hypergiants are so hard to study is they are extremely rare. Conditions are violent on the surface of stars and Eta Carinae is no different. The huge star blows off much of its surface during these violent explosions, possibly blowing off more than 1,000 times the mass of Earth every six months. Imagine a storm so big and strong that each year it would erode 2,000 Earths. Some storms on Earth are strong enough to erode parts of beaches, but this storm would erode 2,000 planets each the size of our own. Eta Carinae's behavior makes it even more difficult for astronomers to determine where the star ends and its solar winds begin. Eta Carinae is expected to blow soon which could make for one of the biggest and most spectacular supernovas.

While there's controversy surrounding which star is actually the biggest this much is agreed upon, the big stars are some of the coolest, literally and figuratively. Eta Carinae is said to burn the brightest of all the stars that we know of. It is in a category of stars called the cool super giants and Eta Carinae burns at some 36,000 degrees Kelvin which would be about 72,440 degrees Fahrenheit. Eta Carinae is actually made up of two stars, the big one is about 90 times the mass of our sun and the smaller one is about 30 times as massive. Together they put out tremendous amounts of light, again thought to be some six million times brighter than the sun. If the big star were moved to the Milky Way its mass would stretch beyond Earth's orbit.

When some of these massive stars explode they can eject enough material to make several new stars. Other times after they blow up they just form a huge dust cloud known as a nebula. When stars of this size spew out material they create something called a solar wind. Some of these winds blow at 7 million miles an hour and can heat gases inside of them to 10 million degrees. They are too big, too hot and too fast to really appreciate and are a long way away, estimated at 7,500 light years out with one light year being 6 trillion miles, which is a safe distance to wonder about these marvels.

The Top Five!

The online publication cosmosup.com has developed a list of what some scientists call the five biggest stars known to date. The Pistol Star, possibly named for its shape is about 25,000 light years away and its mass could be up to 150 times greater than the sun. The Pistol was discovered by the Hubble Space Telescope in 1991. Scientists say the energy released from the Pistol every 20 seconds is equal to what our sun releases in one year. The Pistol Star is about 10 million times brighter than the Sun and much bigger, with is mass taking up all of Earth's orbit than then some.

Next is a star called Mu Cephei and it is considerably closer, about 6,000 light years away. It falls into the category of red super giant stars and might be the largest star that can be seen with the naked eye. The star is unstable, possibly because it is nearing the end of its life and it takes about one billion suns to equal its mass. It also takes about 100,000 suns to equal the brightness of one Mu Cephei.

Next in line is something called VV Cephei A. It is even closer at a distance of about 5,000 light years from Earth. It too belongs to the category of red super giants because its brightness and mass disqualify it from being with the hypergiants. Not a lot is known about VV Cephei A other than it is possibly part of a binary system, which means it has a partner star in its orbit. It size is difficult to measure in part because of its unstable luminosity, possibly caused by the uncertain distance of its partner star. Sometimes when a partner star is too close two stars can seem like one.

VY Canis Majoris, which we've already learned a little about, is also on the top five list. It is as much as 40 times heavier than the sun and some 2000 times larger. VY Canis Majoris is about 5,000 light years away and is about 500,000 times brighter than the sun.

But according to a growing number of experts, R136a1 is thought to be the biggest star. But its status will likely change with new discoveries on the horizon. It is said to be more than 10 million times brighter the sun. Experts tell us that it releases more energy than all of the stars combined in the Orion Nebula. Located about 165,000 light years from Earth it is in something called the Tarantula Nebula which is a cloud that orbits the Milky Way.

Here's a brief summary. Red giants are stars that have burned their hydrogen fuel in the cores and the star begins to collapse under its own gravity. At this point mostly helium is left in the core and that starts to burn even hotter than the hydrogen which makes the star expand and become less dense, with the exception of its core. As the star expands its surface temperature drops and the color on the surface turns to red, hence the name red giant and can be 400 times the size of the sun. After a star has exhausted all of its nuclear fuel it collapses again under its own weight. Gravity creates tremendous pressure and heat strips the electrons from its atoms and it becomes a white dwarf. Electrons can be compressed much closer together than atoms so the white dwarf become extremely heavy and dense. A white dwarf the size of Earth could be more than 350,000 times heavier than our planet.

A Supernova is the term used to describe the death of a large star, one that is at least ten times as big as the sun. The cores in the big stars contract just like the cores of the smaller stars but the temperatures inside the cores of the big stars are much higher, so hot that carbon is fused into magnesium. When the carbon is gone a new cycle begins, there is more expansion and more heat which fuses heavier elements like oxygen and silicon. The process stops at iron because iron requires energy instead of releasing energy and the star collapses for the last time and in sort of a rebound fashion sends out tremendous amounts of light and energy thought to be 100 billion times brighter than the sun. It is this radiation that goes out into the universe that gives life to everything and spawns the term "we are all made of star dust" and "not only are we in the universe but the universe is also in us."

Neutron stars are some of the densest objects in the universe are can be created from the aftermath of a supernova explosion. They can be as heavy as the sun but only as large as a city like Chicago. They might be similar to white dwarfs in that white dwarfs had their electrons stripped from its atoms, and neutron stars have had their neutrons stripped from its atoms and in both processes neutrons and electrons can be compressed together much more tightly than atoms, hence their great weights despite their small sizes. If neutron stars were larger their tremendous gravities might create black holes.

Population II stars are the older ones that fuse hydrogen and helium in their cores, Population I Stars are the younger ones that fuse heavier elements like oxygen and carbon in their cores. Main sequence stars are the most common, making up about 90 percent of the ones in the universe and are characterized by the fact that they fuse hydrogen atoms to helium atoms in the cores like our sun. Binary stars are again two stars that orbit each other.

The space between stars is just that, space, empty and void. According to scientists about 99 percent of the universe is empty which appropriately enough is called space. Who knows how many stars are out there? Astronomers say there could be up to 400 billion stars in the Milky Way and that's just one galaxy out of a possible 100 billion galaxies in the universe.

Stars are also not close together, which can underscore just how big the universe because it houses so many of them with so much room to spare. With the exception of binary stars, which are stars that orbit each other, the closest star is about four light years away which would take more than 60 thousand years to reach traveling in something like the Space Shuttle. Stars also vary in size, some as small as major American city and others too big to comprehend, more than a thousand times larger than the Sun, remember you could fit a million Earths into one Sun. Those small stars can be really dense and heavy, so much so that a grain of its material would weigh 450 tons or as much as fully loaded jumbo jet. Scientists say that every star you see at night in the sky is not only bigger than our sun but it is also brighter. Some of the coolest stars are red, some of the hottest are blue and the biggest stars seem to have the shortest lives because of the faster rates that they burn through their fuels. Seems stars and people have much in common in that one term is used to refer to the entire group but within the entire group is tremendous variation and that's not to mention stars and people are also made of the same basic material.

Too Big to Name!

Some of the biggest formations in the universe don't really have names they're just called structures and so goes the story of something called the Large Quasar Group. There is one of these large structures out there estimated at four billion light years from end to end. Remember light travels six trillion miles in one year, which is another statistic that is too big for human comprehension. But why stop there, here's another fact that's too difficult to fully appreciate, if the Large Quasar Group, known as LQG was in our solar system it would cover the entire solar system at least a billion times over. In fact, not only would the LQG cover the Milky Way but it would also cover the neighboring galaxy in Andromeda and all the space in between those two galaxies.

The LQG is so big that if light traveled from one end to the other end it would take about four and a half billion years, which is Earth's entire life span. These big structures pose big questions like, the basic understanding of the scale of the universe could be called into question. Something so big that it takes light four billion years to cross, threatens to question the cosmological principal which has been widely accepted since the work of Albert Einstein. The cosmological principal is the basic assumption that the universe looks the same size and same shape no matter the angle at which it is viewed.

Quasars are known to lump together and in the process they create these huge structures hence their name Large Quasar Group. Remember quasar is short for quasi-stellar radio source and the LQG is made up of 73 different quasars and its size threatens to force the way scientists look at the upper limits of structures that are allowed in the universe. At some four billion light years across it seems to eclipse what the old structural limit was of 1.2 billion light years. Again scientists cannot explain how it came to this great size but are thankful it is not closer than what it is, about 9 billion light years away.

Some of the problems posed by the great size of the LQG might be best summed up with the abbreviated statement made by Dr. Roger Clowes in the journal Science Daily when he said, the distances and sizes contained in the universe could have been over simplified after the discovery of an LQG and the reality that is left could be more complex than what we can currently understand. Which could be the sophisticated way of saying, the more we know the more we realize we don't know.

Stormy!

The planets in our solar system can be stormy places. Earth's storms would pale in comparison to some of these violent systems on other planets. For example, there is a hexagonal storm raging on the surface of Saturn that has been active for years. It is a storm with six straight sides and a flat appearance. It was first noticed 30 years ago when the probes Voyager 1 and Voyager 11 flew by the planet's gaseous surface. The big storm has been compared to a giant hurricane which has situated itself over Saturn's north pole. The hurricane like storm is so big that it would cover Earth four times over. It is like a jet stream with ferocious winds that spans 20,000 miles. A storm like that has never been seen before in the entire solar system. Imagine a storm so big and strong that you don't ever recall living without it, it's been there your entire life.

Scientists are not quite sure what causes the storm but they have their theories. Since Saturn has no solid land mass, the absence of a solid surface could be part of the explanation. Solid land helps to break up storms so the absence of land mass could be the reason why the storm has not dissipated in three decades, maybe longer. The winds on Saturn could also help explain the storm's geometric shape. Winds on that planet can exceed 1,000 miles an hour. Of course humans could not withstand such a force and would likely be dismembered during the exposure. Scientists think that as that the jet stream pushes its way through a slower moving atmosphere it creates the storm's hexagonal shape.

Astronomers have been watching storms on Saturn for more than one hundred years. One of the first storms observed was in 1876. Since that time, at least six of these big weather systems have erupted over the planet's surface. Traditionally they have been called Great White Spots and each storm is about the size of Earth. They seem to occur every 30 or 40 years and they end up engulfing most of Saturn.

Convection causes more than 15 million storms to occur on Earth every year. Convection is the mixing of warm and cool air that rises and falls in the atmosphere with water condensing and falling as rain during the process. With little or no convection on Saturn, the planet might have only one storm in 30 years, some scientists say only one every one hundred years. So what storms on Saturn lack in numbers they make up for in intensity, again with some of these weather systems engulfing the entire planet and lasting 30 years or even a century.

The atmosphere of Saturn is made up mostly of hydrogen and helium and researchers think that this composition combines to make storms so rare on that planet. Apparently water is heavier than hydrogen or helium which makes it harder for water to rise in the atmosphere and then fall as rain after it is condensed. So this process might explain why the convection process takes so much longer on Saturn.

Saturn is a very cold place with an average temperature of about 270 degrees below zero. These low temperatures could be yet another reason why it takes 30 years or longer for Saturn to generate a storm. So while a hurricane on Earth might last for a week, one on Saturn could well last a century. The huge polar jet stream on Saturn pushes off smaller hurricane like storms, but they are still bigger than the largest hurricane ever recorded on Earth. That would be Hurricane Patricia which developed in 2015 and had wind speeds of more than 200 miles an hour. It lasted for four days and was so big that the storm covered several states.

One can only imagine what kind of lightening is produced by a storm that generates 1,000 mile an hour winds. Saturn's serene appearance belies the fact that these long lasting and high force winds scour its gaseous surface and produce lightning of tremendous strength. Some of the lightning bursts on Saturn are 10,000 times stronger than on Earth, making lightning on Saturn the most powerful in the solar system. The strength of lightning strikes on Uranus and Neptune are under debate. Lightning strikes on Jupiter are estimated to be about 1,000 times stronger than Earth's. Lightning on Earth can produce 10 to one million volts of electricity and reach temperatures six times as hot as the surface of the Sun. An average lightning bolt on Earth has enough power to satisfy all of the energy needs of a small town for a day. Saturn's lightning can be a million times stronger which might be a shame considering it has no small towns to power.

Not only does Saturn check in with the strongest lightning but it apparently also has the longest lasting storm detected in the solar system. On Earth we have Tornado Alley, Saturn has Storm Alley. Storm Alley is situated on Saturn's southern hemisphere and this is where many of the lightning strikes are thought to occur. None have ever been seen, in part because of the glare and reflection of Saturn's rings, but scientists are certain the lightning strikes have occurred because of the radio waves left behind. Scientists are not sure what causes lightning on Saturn but one theory is the strikes are created deep in the planet's gaseous atmosphere where light from lightning is much harder to detect. Seems Saturn's behavior is every bit as strange as its appearance which only makes the planet more appealing and interesting and dangerous.

What's on Uranus?

Here is a word to the wise, be careful how you state this next fact so as to avoid: "There's a big storm on Uranus." which would be both correct and incorrect at the same time. Yes, there is a big storm on the giant gas planet far out in our solar storm, but there are no weather patterns detected on one's buttocks. Making matters more difficult, I would not want to imagine the odor of the storm on Uranus when you consider Uranus is made up of a lot of methane gas which is quite similar to the methane gas found in cases of human flatulence, better known as the common fart. Technically speaking methane only makes up only about seven percent of the human fart and apparently sulfur is to blame for its offensive odor but the fact remains the same chemical is present on that planet and in our gas.

Fact can sometimes be stranger than fiction. So are we to assume that Uranus is a cold, dark, smelly place? That was an example of the potential embarrassment that can be caused by the planet's name in a particular context. What we do know is Uranus is the butt of cosmic humor. Anyway, the big methane storm on Uranus was first spotted in 2014 in the Northern hemisphere of the planet. It is an area of Uranus that has not seen sun light in some 20 years, and according to one theory the planet is entering its spring and the slight warming effect might have contributed to the storm's creation.

Remember seasons can last for 20 years on Uranus because of its near 99-degree tilt. Some seven years after it had entered its summer in the mid 2000's there was another storm discovered on Uranus. Again, because of its radical tilt, the summer sun shone directly on the Uranus's equator for decades. Because Uranus has no real internal heat, it is thought that the Sun's heat, albeit very weak at that great distance, was powering these storms. Since then more than a half a dozen storms have been observed on Uranus, some even by amateur astronomers. All of these storms were big, the recorded size of one weather system was 5,500 miles across which would be enough to cover the United States twice if you traveled the country in a straight line.

For some reason 2014 seemed to be the stormiest year on Uranus which came seven years after the planet's equinox, which actually could have answered the question. For our purposes an equinox can be defined as a period of time in its yearly orbit where a planet receives direct light from the sun. So that direct sunlight could have caused the stormiest year! During this time a total of eight storms were spotted on Uranus and all of them were near the planet's North pole, which is similar to Earth's equator in that both areas receive direct sunlight during their equinox periods.

Astronomers don't know what causes this but suspect Uranus's characteristics might hold part of the answer. Remember it takes Uranus 84 years to complete one orbit of the Sun and in that time one of its poles is exposed to 42 years of darkness. But during the equinox of 2007 sun light reached both of its poles in equal amounts which could have caused the increase in storms across the surface. Lows on Uranus can reach 370 degrees below zero and its winds can blow more than 550 miles an hour, combine this with sun light, however weak at that distance and it might be all that is needed in such a volatile environment to create some of the storms that astronomers have recently observed.

Scientists also think there is lightning on Uranus, this after they heard radio signals of lightning bolts. It was the same kind of static booms heard on a standard radio during an electrical storm on Earth. The bolts of lightning came from the upper atmosphere of Uranus and are said to be much stronger than the ones on Earth. So while Uranus might be may things, it even was once called the "Boring Planet" but really it is anything but with its awkward name, mysterious behavior, massive storms and offensive odor.

What Weather?

The closest planet to the Sun is Mercury which hardly has any atmosphere at all. So it does not have weather in the form of storms like on Earth, there are no clouds, nor wind or rain or even dust storms but it is anything but boring. Its surface temperature can reach almost 800 degrees in the day and plummet almost a thousand degrees in the other direction at night to 279 degrees below zero. The massive temperature shift is due to the fact that Mercury has no atmosphere to trap heat at night.

Apparently Mercury has magnetic storms that could take the place of storms generated by weather. The planet's magnetic field releases energy in violent disturbances which are called sub storms. These sub storms are stronger and more violent than the storms on Earth. Mercury's magnetic field is somewhat of a surprise to researchers because the planet spins on its axis at a very slow rate making a completion once every 58 days. So a day on Mercury lasts nearly two months. Earth spins on its axis once every 24 hours. Scientist think that the faster a planet spins, the greater its magnetic field which is why they were surprised to find a magnetic field so strong on Mercury.

Mercury has no seasons, just really hot and really cold, in part because of its close proximity to the Sun and something called a tidal lock. The small planet is tidally locked in place in its orbit around the sun which means it does not spin on its axis. It is quite common for planets too close to their stars or moons not to spin on their axes so they instead show the same face to their host planets or stars, as does Earth's Moon.

Since there is not much weather in the traditional sense to talk about on Mercury we will focus more on its climate, which is almost totally dictated by its close proximity to the Sun. During the day Mercury can receive up to 10 times the amount of Sun that we get on Earth. If humans were exposed to that kind of heat generated from the sun, we just might explode as in spontaneous combustion. Spontaneous combustion is defined as the ignition of organic material, including human bodies without apparent cause, typically through heat generated internally by rapid oxidation or loss of electrons. No science fiction here, what do you think would happen if you climbed inside an oven set on 800 degrees, which is about the equivalent temperature in the aforementioned scenario?

When Mercury is tilted on its axis its southern hemisphere gets hotter summers and cooler winters which produces some of the most extreme weather fluctuations in the solar system. Besides being the closest planet to the Sun, Mercury is also the smallest planet in the solar system. Mercury is just a little bigger than the Moon. Another interesting fact is how such a small planet could survive such a big asteroid impact. That big collision was thought to have happened some 4 billion years ago and it left a crater the size of Texas on its surface. That's almost a thousand miles across and some 60 miles deep.

Besides being the smallest planet on the block it is also the fastest. Mercury completes its orbit around in the Sun in 88 days traveling at 112,000 miles an hour. Earth is almost twice as slow, traveling at some 67,000 miles an hour and it takes 365 days to complete one of its orbits. It is often said that Mercury is home to ice cubes in hell and that could be a reference to the possibility that ice could be buried in its craters. The small, speedy planet also has a tail, possibly due to its close proximity to the sun. Remember the Sun pummels Mercury with a constant barrage of radioactive particles, many of which are believed to make up this tail as they are get shot out from the Sun and are ricocheted off of Mercury's surface. The discovery of Mercury's tail was made in 2004 by a probe that flew as close to the hot planet as it could without melting, an astonishing 142 miles from its surface. It will be difficult to learn more about Mercury because it is so hot and so near the Sun.

He Did What!

You might not have heard of some of the people who lead the space race with their deaf defying feats, unfortunately some of those feats did not defy death but all were performed in the name of science. One person that comes to mind is Yuri Gagarin. The man was extreme long before there were extreme sports. Some say Mr. Gagarin was the first astronaut because he was the first person to actually travel in outer space. In the spring of 1961 Russian Cosmonaut and Pilot Yuri Gagarin was set to help his country win the space race with a feat that had never before been accomplished, becoming the first man in space with the launch of Vostok program.

He reached an altitude of 203 miles and traveled at 17,500 miles an hour on his way into the record books. He accomplished all of this with technology that had not been tested and was at times extremely dangerous. Mr. Gagarin had no choice but to take part in the mission as he was "selected" by the Soviet Government. A few minutes into the launch one of floor panels dropped away to reveal the view below which was designed so the cosmonaut could orient the craft with the sun and the horizon line. It was a window on the world never before seen by a human. No one could have blamed Mr. Gagarin if he lost his nerve and needed a relief bag but that was not the case for this highly trained, talented and courageous cosmonaut.

One of the problems with the space race was the proverbial breaks had not yet been invented. In short, the Soviets had a plan to get a man into space but had no real plan to return him to Earth. So part of what they had in mind was an automatic ejection where Yuri Gagarin would be thrown from the craft before it smashed into Earth. His ship was built is the shape of a sphere in order to deal with the changes in the center of gravity and make it as comfortable as possible for one person to orient himself. The craft had no thrusters to slow down so the idea was for Yuri to eject from the speeding vessel at an altitude of four miles. On the way down Yuri Gagarin had to deal with 8 Gs of deceleration, which is the equivalent of eight people of the same weight sitting on top of him.

All this was done 50 years before somebody called Felix Baumgartner. Mr. Baumgartner was the high altitude parachuting sky diver who jumped from a distance of about 20 miles in 2012 which at the time gave him the world record for such accomplishments. Mr. Baumgartner could choose when he would jump, Mr. Gagarin could not because he was "selected" for the mission. Mr. Baumgartner had the benefit of modern technology. Mr. Gagarin did not. Still Mr. Gagarin managed not to kill himself and returned to Earth in one piece.

One month later Alan Shepard would become the first American in space but Yuri Gagarin was the person there, period. Mr. Gagarin became an instant intentional celebrity. He was a national treasure, too valuable for the Soviets to send up again. At 5'2" tall, Yuri became a giant figure of Soviet success. He was just 27 years old, but before he could return to space he was killed in a training accident at the tender age of 34. He was cremated and his ashes were buried inside the Walls of the Kremlin on Red Square. He remained a hero even after the Soviet Union was dismantled and his statues were preserved while monuments to other Soviet leaders were torn town.

Can't Make This Stuff Up!

A trio of astronauts could have been called Three Musketeers in which "all for one and one for all" not only was their motto but was the reason for their success. They are Pierre J. Thuot, Thomas Akers and Richard J. Hieb. They made history in the spring on 1992 when they rescued a stranded satellite with their bare hands. Well their hands were covered by gloves that protected them from the dangers of space but their achievements should be no less diminished by the technicality of their protective gear.

It took about four hours to retrieve the faulty satellite from an orbit estimated to be about 22,000 miles above Earth. The astronauts were inside the shuttle and the satellite they needed to maneuver was about 17 feet tall. The original device intended to retrieve the satellite didn't work, even though it cost some 7 million dollars. So the three astronauts positioned themselves outside of the shuttle and inched closer to the satellite. The new device that was going to retrieve the satellite was the astronaut's hands, which was about as low tech as you could get in the high tech world of space exploration.

One astronaut was perched on the steps of the shuttle's cargo bay while the other two were in similar precariously perched positions. Each moment brought them closer and closer until one of them said "I got it." Cheers then broke out on the ground below at mission control. According to some reports all three astronauts grabbed the multimillion dollar satellite at the same time, the "all for one and one for all mentality" apparently saved NASA and the United States considerable embarrassment because it looked as if the mission would end in failure. In fact, foreign journalists admitted to following the story because of that possible embarrassment to America's space program.

To high light some of the dangers of the trip, just consider some of the many things that could have gone wrong. The several ton satellite could have crashed into the shuttle. Even minimal damage to the shuttle could have been a disaster because upon reentry the slightest of harm could have been fatal. What happened it part of the jagged satellite contacted the astronauts? It could have easily sliced through their protective gear and released their oxygen. So the astronauts had five layers of protective material on their gloves. Engineers spent two days studying designs and data to make sure no sharp edges could puncture the astronaut's suits, still no guarantees could be offered.

All this came at a time when the American media became bored with the many shuttle flights which at that time seem to be almost routine. But this time the world seemed to be watching and what people were seeing was not scripted, this was live and could have fatal consequences. The three men positioned themselves around the satellite like the three legs of a tripod and in the process of defying death they made history. Just like the Three Musketeers might have done, one astronaut said "let's do it" and then another said "got it" and that was it. Their success was a milestone because until then neither the United States or Russia had three people outside a space ship at one time, made possible in part by the mentality of the Three Musketeers.

Space exploration is chalked full of amazing, interesting and funny stories and this next tale has all three elements. Even though NASA has made tremendous accomplishments in space exploration it might still have room for improvement when it comes to the management of human waste, which is an introduction to the case of frozen urine stuck to the side of one its shuttles during orbit. This problem presented itself in 1984 on board the shuttle Discovery. The shuttle's waste dump system malfunctioned and was no longer getting rid of the astronaut's urine. A short while later, a large urine icicle formed on the outside of the shuttle. It didn't take long for the offensive structure to grow to 30 pounds, which could be the weight of an average two-year-old child. This might seem funny if it did not have the potential to be fatal. Problem was that big urine icicle could have broken off during reentry and hit part of the shuttle which again could have had deadly consequences for the entire crew.

A remedy to such a problem was not in the NASA training manual so the astronauts were on their own and made things up as the went. Their first attempt was to melt that hideous growth right off the side of the shuttle. That option brought quite a bit of optimism because they were going to use the all-powerful sun as a heat source. But it didn't work. For three long days the shuttle was angled toward the sun. Needless to say crew members were discouraged to find out it didn't work. After 72 hours of direct exposure to the sun, its rays did not even make a dent in the 30-pound urine prong that was now part of the shuttle. That result really should not have been a surprise considering the average temperature outside of a space shuttle is well below 200 degrees below zero. Sounds like the urine was frozen more solid than rock. Using a spacewalk didn't seem like an option because of the dangers involved, let alone selecting a volunteer to do the dirty work. So another option was selected. They would use the shuttle's grabber arms to break apart that frozen problem that had attached itself to the craft. The craft was America's pride and joy and part of it was covered in urine. The arm was able to exert enough force and sent the frozen problem into that big urinal of the great beyond. No doubt all astronauts have stories they tell to the children and some stories they keep to themselves, wonder which category this one would fall into.

When's the Next Time?

When will the next asteroid hit Earth? Problem is no one knows but according to NASA it won't be anytime soon. Another problem is nobody saw the asteroid coming that could have wiped out parts of Siberia when it exploded above its sky in 2013. Despite the fact that many eyes are trained on different parts of the solar system to detect incoming asteroids and comets, the one spot we cannot look is in the direction of the sun and that's the blind spot from where the asteroid in 2013 just suddenly appeared. It came out of nowhere, with no notice nor advanced warning. Predicting the next asteroid strike might be similar to trying to predict the future in that trends and past occurrences are considered but you won't get killed because of an incorrect prediction about the future but you certainly could if you're wrong about the next asteroid impact.

If it's any consolation, NASA states there are no foreseeable asteroid dangers for at least the next one hundred years or so, of course barring those that come from that blind spot. Astronomers work at a number of locations including the Near Earth Object Office Program to hunt for these potential planet killers. They do this by using a number of telescopes and other high tech equipment. They say with an element of relief that don't detect anything out there in the near future that threatens to crash into Earth. Of course our planet is bombarded with lots of asteroids and other space debris on a daily basis but these are too small to do any real damage and usually burn up in the atmosphere. But it might be difficult to find comfort in those predictions as long as that blind spot remains part of the equation.

The planet seems to be safe from asteroid impacts during our lifetimes, what about the lifetimes of our children or even grandchildren. Time to introduce something called 1950 AD. It has been called a giant space rock, even by asteroid standards and it could hit Earth in 2880. There are also two other threatening asteroids that might hit as early as 2175 and 2185. The one most likely to hit is called 2009 FD. It is nearly a half of kilometer in size and depending on where it could hit would determine how much damage it would cause and how many lives would be destroyed. We're told if 2009 FD hit in the ocean, a giant tsunami would ensue followed by a large loss of human life. It if hit in a populated area it could wipe out entire cities, maybe even countries. The chances of a 2009 FD impact are about one in 340 which are about the same odds as a child being born with debilitating condition. 2009 FD was discovered in 2009 and making matters worse it has more than one chance of striking Earth. Astronomers tell us that has 2009 FD flies by there are actually several times where it could smack into Earth including in 2185. The size of this asteroid would enable it to break directly through Earth's atmosphere and impact the surface with more than 17,000 times the energy of the Hiroshima bomb. It would also leave an impact crater some four miles in diameter and its climate effects could last for years or even decades.

And where there is one there is usually another and this is the case with something called 1999 RQ36. It is a little bit bigger than 2009 FD and could hit Earth in the year 2175. Not only is it a little bit larger but it also has a much bigger chance of hitting Earth when you consider it has almost 80 opportunities of striking our planet as it flies by during the years of 2175 to 2199. Again this is not in your life time but possibly in the life time of your child or grandchild's. So 2009 FD has several chances of hitting Earth when it flies by in 2185 to 2196 and the slightly bigger 1999 RQ36 has 78 chances of hitting us when it flies by during the years of 2175 to 2199. Long story short, in one particular fly by 1999 RQ36 has a one in 1000 chance of hitting Earth which are about the same odds as a person getting struck by a bus as that person crosses the street. During all of its fly byes or any one of its fly byes, 1999 RQ36 has a 1 in 2,700 chance of hitting Earth with the year 2175 as the earliest dooms day date for any kind of impact. Of course the blind side is not factored into the equation.

The biggest and possibly most devastating of the bit three space rocks is 1950 DA. It was discovered 65 years ago and is about a mile in diameter traveling at more than 25,000 miles an hour and that blistering speed can make all the difference. NASA engineers calculate it has one realistic chance of impacting earth in the year 2880 and that chance of impact is about 1 in 20,000. Those are about the same odds as getting run over by a car. So while the threats posed by the big three are a long way off, in the next century, if we don't beat the odds the next century could be the last.

We are told that NASA does its best to monitor these potentially catastrophic asteroids and follows all of them that have orbits with in several million miles of Earth. Those orbits would include crossing over the orbital paths of Mars, Venus and of course Earth. Adding one problem to another, NASA or anything other organization for that matter, has no current plan in place to intercept these big space rocks. Apparently the best we can do is to predict that the punch is coming but we have no plan of getting out of the way.

At best, we would need 60 to 100 years advanced warning in order to try and change the orbits of these asteroids, the sticking point being, we arguably don't have the technology required to do the job. There are an estimated several million asteroids bigger than 100 meters that are in our solar system and NASA is aware of only 13,000 of them. So this appears to be a cosmic game of chance and the future of our planet could be at stake.

Methods of deflecting or diverting an asteroid are only in the planning stages but some of them would include trying to attach some sort of solar sail on the surface of an asteroid in an effort to change its path. Another possibility might be to coat or dust part of an asteroid with a substance that would interact with sun light so as to change its direction. But again all of these possible methods asteroid deflection requires advanced warning of at least 60 years.

Here's a quick recap of what might happen during and after an asteroid impact. Of course much depends on its speed and size, but either way the results would not be good. If the asteroid was larger than a mile and traveling at 30,000 miles an hour it might wipe out most of the life on the continent. If the asteroid was smaller, say the size of a common three-bedroom house, it would do significant damage to any city and probably kill millions of people. It would flatten steel and concrete buildings within a half mile from the impact site and level wooden structures up to two miles away.

Obviously the big asteroids are most problematic, one a mile wide that directly hit New York City might level everything in the general vicinity from the Washington D.C. to Boston, MA. The dust and ash kicked up in the atmosphere would have a global impact and probably choke the life out of most living things on the planet. Again we are told the planet is safe for the remainder of his century, but as always asteroids from the blind side are excluded. So the next century could user in a new era, similar to the one that ushered out the dinosaurs.

What might be lost in the amazement of asteroids is they are like bombs when they strike Earth, only worse because they are more powerful. Many experts are convinced they are more destructive than nuclear weapons when they hit. The asteroid that broke up in the sky over Siberia in 2013 was equivalent to 30 atom bombs. The atom bomb dropped on Hiroshima instantly leveled everything within a mile, an estimated 60 thousand buildings were destroyed and 80 thousand people were killed in the initial blast. An explosion 30 times stronger is hard to comprehend but that was the amount of energy released when the asteroid in 2013 broke up above Siberia. It was about 15 miles above Earth when it exploded, large scale death and destruction were certain had it impacted the ground.

It traveled at 40 thousand miles an hour and became as hot as 3,600 degrees which is about seven times hotter than your oven's maximum temperature setting. Making matters worse is the fact than an asteroid's path can be easily changed, even light from the sun can alter their orbits which casts doubts on when and where they are predicted to hit, which could be a moot point because this asteroid came from the blind side.

This was not a big asteroid, small by galactic standards maybe the size of a du-plex apartment building but it what it lacked in size it makes up for in speed which illustrates the new threat that smaller asteroids are much more dangerous than previously thought. Smaller asteroids are also harder to detect made this way by their dark color pitted against the pitch blackness of space. One such asteroid is DA14, scientist had tracked it for more than a year and were certain we were safe because it would not hit Earth, problem was on the very same day in 2013, Earth was hit but by another asteroid which injured more than a thousand people when it came from the blind side and exploded in the sky over Siberia.

The 2013 Siberian asteroid could have easily been much more destructive if not for just a few simple changes in its composition and trajectory. This asteroid was old and battered with cracks and fractures. Comprised mostly of rock and only about ten percent metal, made it more likely to break up in the atmosphere as opposed to making it down to the surface. Reason being, metal is much more durable than rock and more likely to survive reentry. The asteroid's trajectory was also beneficial, it came in at a slant instead of straight down, which meant it passed through more atmospheric friction that caused it to break apart as opposed to coming straight down with less friction and a greater chance of hitting Earth in one piece. On a technical point, scientists say the asteroid didn't hit us but instead we hit it, as Earth traveled in its orbit at 67,000 miles an hour, we sort of rolled right over the small asteroid.

They have hit us before and we're told they will hit us again. Scientists have complied a list of the ten largest asteroids that have already hit Earth. First on the list is the one that left behind the Vredefort Crater in South Africa. That asteroid hit some 2 billion years ago and left behind the largest impact crater to date which measures some 118 miles in radius. The Chicxulube asteroid was the one that killed the dinosaurs and only ranks eighth on the top ten list. The Chicxulube asteroid struck some 65 million years ago and left behind a crater 106 miles in radius. That should illustrate that there have been several asteroid impacts in the past, large enough to cause global extinction which begs the questions how much borrowed time are we living on.

Long Time!

If slow and steady wins the race than Valeri Polyakov is the clear victor. He's the man who has spent the most consecutive time in space without a break, living on the Mir Space Station for 14 months which is 437 days. Considering the size of the station it could be compared to spending more than a year inside a one-bedroom apartment. Except the apartment would probably be roomier when you realize that the Mir Station was actually a laboratory designed for scientific observation and completely full of lots of high tech equipment.

Besides being a world record holder, Mr. Polyakov is also a medical doctor. In fact, he volunteered for the mission to prove his point that humans could travel to Mars. He credited his success to a strong and vigorous work out program and as a physician he was well aware of the effects the trip and its micro gravity would have on the human body. Even though his trip was a success, the mission got off to a rough start. The space craft that transported him to the Mir station actually hit the station but fortunately no major damage was done to either craft, sounds like the equivalent of a fender bender in space.

When he returned to Earth he was described as looking big and strong and he told on lookers "he could wrestle a bear." Contrary to the beliefs of his medical profession, he also had a cigarette when he landed. Taking about cabin fever, Mr. Polyakov spent 437 days in space from January of 1994 to March of the next year. His accomplishment raises the question of how long can humans stay in space. Apparently Valeri Polyakov never developed a case of space sickness, or at least there were no reporters about it. On Earth, humans get their sense of balance with their eyes and ears. In space only the eyes are used which can lead to space sickness, which is a fancy term for throwing up. About half of the people who travel in space contract the illness which is defined as general nausea brought about by the change in G-forces and spatial orientation.

Apparently Mr. Polyakov also had no problem sleeping in Earth's low orbit which extends from sea level to about 1,242 miles up. In the low Earth orbit the Sun rises and sets every 90 minutes which can disrupt the natural cycle of sleep for many astronauts. That's one reason why mission control will play music in an attempt to keep astronauts grounded on a regular 24-hour cycle. In fact, Valeri Polyakov might have actually slept better in space, this after a study that showed astronauts who snore on Earth sleep silently in space. Apparently zero gravity plays a role in this phenomena as the tongue and jaw are not pulled back toward the throat by gravity thereby creating less obstruction in the airways.

Mr. Polyakov was also exposed to significant amounts of cosmic radiation which have afflicted a significant number of astronauts with cataracts. When astronauts gaze out of their windows they are privileged to extraordinary views and also to extraordinary dangers. More than 30 astronauts have reported seeing bright flashes of light inside their eyes. Apparently space radiation actually rips through their eyes like the sub atomic particles that they are. When these particles contact the retina it sends off false signals that the brain thinks are some form of light. This might help to explain why some astronauts in space saw bright flashes outside of their capsules and when they pulled the blinds down over the windows they still saw the bright lights, so they closed their eyes and yet they still saw those same lights. No doubt after spending 437 days in space Mr. Polyakov logged ample time enjoying the view from his bay window but no word on if he developed cataracts.

It would only stand to reason that there was a good chance that Mr. Polyakov also spent more than a year without a shower. Showers just don't work well in zero gravity so astronauts stay clean by taking sponge baths. The Mir station was equipped with a shower but still most astronauts prefer to take sponge baths.

Mir was Russia's first great space station and its name can be translated to mean "world", "peace" or even "village", an admirable name given its definition and also a good indication of the international contribution that would follow in future space stations. Mir was the fore runner to the International Space Station which would take its place as the largest satellite to orbit our planet. Mir was constructed using a series of modules, its core module was about 43 feet long and this was where the astronauts conducted their work and other vital activities.

When the International Space Station became fully functional the Mir station was abandoned. In 2001 Mir was taken out of orbit and crashed back down to Earth in a fireball somewhere over Fiji. The Russian government was worried that Mir might crash into a populated area so it purchased one massive insurance policy that it fortunately never needed to use. Mir's total cost was estimated to be 4.2 billion dollars and it stayed in orbit an amazing 15 years. In that time, it was visited by 104 people from 12 different nations and it made 86,000 orbits of Earth. All that's left now are the memories of the machine that made history and the talented man and women who dared to dream. When it comes to traveling to Mars, one of those dreamers is Mr. Polyakov who confidently said after spending the most time in space of any human, that the trip can be done.

Still the Fastest!

One might think that the speed at which technology advances that records set in the 1960s would have been broken several times over by now but that's not the case with the title of the fastest humans. That record belongs to the crew of Apollo 10. In May of 1969 its three-man crew achieved the highest speed ever reached by humans, blasting through space at 24,791 miles an hour, fast enough to keep pace with an asteroid.

Apollo 10 was seen as the last dress rehearsal for the moon landing as it was designed to travel around the moon without actually landing on it. It came within nine miles of the lunar surface, a razor's edge in galactic terms. Much has been made of its top speed which was not that much faster than other missions of than time but there's no definitive explanations how it was able to achieve that top speed.

The time of the month might have been one factor. The moon does not orbit in a circular pattern so the distance varies which could affect the speed of the craft because the gravitation forces on the space craft would also fluctuate. The mass of Apollo Ten might be another factor. Remember it did not land on the Moon so it was lighter because it did not have to bring back moon rocks to Earth for research. Apparently there were also a number of course corrections made on the way back to Earth which could have also increased its speed.

Despite all of its technological innovation Apollo 10's procedure for the handling of human waste left much to be desired. That method meant astronauts evacuated their bowels into a plastic bag and the bag was then tapped to their legs, contents and all. As Murphy, of Murphy's Law predicted, "anything that can go wrong will go wrong." Long story short, astronauts had to deal with what they called "floating turds." Suddenly being an astronaut doesn't sound so glamorous. Fortunately, the astronauts were trained professionals and they just joked about the messy problem, some blaming others for the "floating turds" while others would say things like "I can neither claim it nor disclaim it." The irony here is some of the smartest people of our generation who were supported by the latest, most sophisticated technology still fared no better, or maybe even worse than our ancestors who lived in caves when it came to disposing of their waste.

Super What!

A supernova is a star at the end of its life and suddenly increases its brightness because of a massive explosion that blows off most of its mass into space. They can be so bright that they outshine an entire galaxy. Remember a galaxy has millions, maybe billions of stars and just one of them is so bright that we can't even look at it.

There are two kinds of supernovas. The first is a binary system where two stars orbit each other and the bigger one takes mass from the smaller one which makes the bigger one eventually explode. The second type happens when a large star is at the end of its life and it collapses and then explodes. Or put another way, the star runs out of fuel, collapses under its own weight and then rebounds in one big explosion. Stars have to be really big to go supernova, our sun does not have enough mass to go out with such a big bang. Suffice it to say we don't want to be anywhere near the home galaxy when these big stars blow.

One that comes to mind is Betelgeuse and scientists say with certainty that this star will go supernova but they don't know when. Betelgeuse is massive, maybe a thousand times larger than the sun. Its home is in the Orion Constellation. That is a long way, some estimates put it at 12 million light years while others put it as close as 430 light years. Remember a light year is six trillion miles. Be glad it's all those light years away, if it were close we'd be exposed to an explosion at least 100 trillion times stronger than the Hiroshima bomb. That's a one followed by twelve zeros 1,000,000,000,000 and it might even be twice that strong. But by the time that tremendous energy reaches

Earth it will have fizzled all the way down to a cosmic light show, albeit probably the most spectacular one ever seen.

No one knows definitively when Betelgeuse will explode but it is thought to be soon according to galactic standards, maybe with in a million years or maybe next week. When it does blow it will not fade away but rather burn out brighter than we could imagine. Betelgeuse's luminosity is expected to brighten the sky, both day and night. It might be as bright as a full moon and its light show could last for days or months, but scientists say it won't be bright enough to be seen as a second sun in the sky.

Again Earth is too far away to be damaged when Betelgeuse blows. According to many astrophysicists Earth would have to be about 50 light years from a supernova to threaten life on the planet. Betelgeuse is thought to be at least ten times that distance from us. If our sun and Betelgeuse were side by side, Betelgeuse would be about 10,000 times as bright. But brightness does not always transfer to heat because our sun is almost twice as hot at Betelgeuse even though the sun is about a thousand times smaller. So where is the big Red Giant known as Betelgeuse? If the Orion Constellation is shaped like a hunter, Betelgeuse would be sitting on the hunter's lower shoulder. The best viewing would be in the evening months of January and February with Betelgeuse rising around sun set.

According to a commonly accepted theory, if Betelgeuse exploded tomorrow we might not realize it but people five generations into the future might. Reason being, if Betelgeuse is 430 light years away, it would then take 430 years for its light to reach us. Even though Betelgeuse is one of the brightest stars in the night sky it would not matter, because light travels at a constant speed no matter how bright.

So while Betelgeuse is big and strong enough to destroy several planets when it goes supernova, it is far enough away that it won't threaten life on Earth, which might enhance our appreciation of the Red Giant with respect that is amplified from a safe distance.

Wow!

The mystery of the WOW signal has been around for decades and has still not been solved. WOW was the name given to a strange radio signal in Ohio during the summer of 1977. It was a strong narrow band signal detected by researchers at the Perkins Observatory. The signal appeared to have some the characteristics of something that was not from Earth or even from the solar system. It was immediately seen as some sort of possible alien communication which came from the direction of the Constellation Sagittarius.

Astronomers were not sure what kind of signal to expect from interstellar space but they were optimistic that this was worth the while in taking subsequent looks. The signal only lasted for 72 seconds and for years afterward researchers and astronomers tried in vain to relocate it. The researcher who observed the signal circled it on a computer printout and wrote the word ''Wow!'' next to it which is how the term came about. The printed signal was a vertical column with the alphanumerical sequence of "6EQUJ5" again with the word "Wow!" written next to it and to this day the mystery still remains.

Since scientists don't know what the signal was they were forced to speculate. It might have been something called at Interstellar Scintillation which is explained as something similar to a twinkling effect in the atmosphere. That's only a partial explanation at best. The researcher who originally received the signal doubted it was extraterrestrial in origin because he said researchers should have been able to find it again. Instead he suggested that it most likely was a signal from Earth that might have been deflected by some debris in space orbiting the planet. We'll find out later that there's a lot of space debris up there.

Despite any evidence to the contrary many people wonder was the radio signal a message in cryptic form from an alien civilization trying to communicate with us. At this point the imagination tends to supersede what little facts have emerged. There is further speculation that a possible alien civilization would use a portion of the radio spectrum that mimics hydrogen. Hydrogen is the most common element in the universe and would be an ideal mode for interstellar communication.

Remember interstellar space is about as remote as one can get, occupying the desolate area between galaxies. It is so bleak and barren that if one cubic foot of atmosphere at sea level contain a billion atoms that same area in interstellar space might only have one atom. So again using a mode of communication based on hydrogen, because of its abundance might have a better chance of making it through the atmosphere or across the vast and bleak areas of interstellar space.

Adding to the mystery is the message might not have been a message at all, or if it was, it might not have been intended for us, which might be why we can't understand it. The observatory in Ohio did not actually receive the message it was more like a signal that was intercepted. Each letter or number on that chart represents about 10 seconds of silence where researchers just listened.

The scale went from 1 to 9 and from A to Z and lasted for little more than a minute. Researchers returned to the area more than 50 times but to no avail. The signal was never found again. Problem is we might never know what the signal was or from where it came. Such questions as, was that a message from alien friends or alien foes we might never know? Was it a signal from Earth that got deflected by space debris? More than 30 years after the "Wow" signal was first detected, scientist sent out more messages in the direction from where the original one came and once again there was no response.

As any good scientist will tell you, "always follow the evidence and when there is none, reserve judgment" but applying that philosophy to the Wow occurrence does not make for an interesting story. There is no clear evidence to suggest that an alien civilization was trying to contact us. Since science has never found any evidence of an alien civilization we must rely on what is more probable that not.

Earlier we mentioned that there is a lot of space debris up there. Consider the fact there are far more than 20,000 manmade pieces of space debris larger than four inches still in Earth's orbit from all of the rocket and satellite launches made over the last 60 years, it seems more probable than not that the Wow Signal came from Earth and was deflected off one of those many bits of debris. The more probable than not method makes the most sense but again it does not make for the most interesting story.

3 Short Days on the Moon!

Did the Apollo 17 astronauts know something that most of us did not when they blasted off to the moon? Did they know their trip would likely be the last lunar venture for the foreseeable future and was that why they stayed so long on the Moon? We may never know the answers but we do know that crew spent the most time humans have spent on the moon since that visit.

One possible answer to that question might have been a resounding yes, maybe NASA's top brass knew in advance that they would not be back to the moon for a while and that was the reason why it decided to send a scientist up there. Until that point no scientist had been to the moon. So the decision was made to send a geologist to the moon, which for a geologist could have been like a trip to heaven.

Geologists usually spend their entire careers studying the composition of Earth, so what better scientist to send to the moon where they would be able to study the moon's surface and rocks which are similar to Earth's and promised to tell us so much about the early development of our planet. The geologist passed the rigorous astronaut training and spent three days on the lunar surface, digging and scratching and poking and collecting.

Beginning in December of 1972 the three men spent a little more than 3 days on the moon. That's almost 72 hours and in that time they managed to perform three space walks, which lasted more than 22 hours. That mission also set other records: the longest moon walks, the largest lunar sample taken and returned to Earth and the longest time in orbit around the moon. Nearly all of the astronaut's time was spent working, taking samples of the moon's surface, deploying scientific instruments, taking measurements and pictures and in general conducting as much research as was possible in that three-day period.

One of the experiments was to try and determine if there was anything to the bright flashes of light that astronauts had complained of during earlier trips. The flashes of light were thought to be caused by cosmic radiation that might cause eye damage in the form of cataracts. So during the Apollo 17 mission crew members carried experimental mice on board. The mice had implants in their skulls to detect if they were effected by radiation, they later developed lesions but they were not caused by radiation. No damage was also found in the retinae of the mice, encouraging news to the astronauts who like most people are concerned about their eye sight.

Their first moon walk was conducted four hours after Apollo 17 touched down which included a ride on the lunar vehicle. During one of the later walks the crew collected 148 pounds of moon rocks that were taken back to Earth. Some of the rocks and dust smelled bad, which became apparent when the astronauts returned to their craft. The foul smelling dust was also hard to contain and found its way into many parts of the capsule. One explanation why the moon dust smelled badly is the moon is more than four billion years old and that was the first time that lunar dust came into contact with oxygen which made for the unique odor. That might be comparable to trying to air out an old pair of tennis shoes, the shoes don't smell until you disturb them. Some astronauts said the moon dust smelled like gun powder, no explaining that because the moon's surface and gun powder are not chemically similar and still no word on what it might have tasted like.

So why did the Apollo missions end with 17? There are conspiracy theories that make for a good Hollywood movie. The movie Apollo 18 was based on the premise that the craft actually landed on the moon in 1974 but never returned and as a result NASA stopped all future moon explorations.

One NASA official called the conspiracy theory pure baloney. In fact, three more lunar missions were planned, 18, 19 and 20. Apparently the Apollo missions ended because of budget constraints and that the missions had accomplished all they intended. NASA had apparently proven to the world that the United States was the leader in space exploration, coupled with the fact that congress was reluctant to allocate more funding for the project signaled that it was an appropriate time to leave that program and look toward future endeavors. One of those projects included the costly trips to Sky Lab.

Our last visit to the moon was in lunar module called the Challenger, it and the crew covered more than 18 miles of the moon's surface. Their work lasted about three days, which included riding on a lunar vehicle that could carry up to two astronauts a top speed of about 10 miles an hour. Almost 150 pounds of moon rocks were gathered in part possibly because the first geologist was on board who might not have been able to contain his enthusiasm for collecting so many rocks. Or according to theorists the astronauts realized this was the last trip to the moon for a while so they collected as much lunar material as they could in their short three-day period. The year was 1972, the last time we set foot on the moon.

The Old Cloud Spawns a New Theory!

While it is commonly thought that planets are formed from nebula, there is a new planetary formation theory that accepts the nebula concept but moves in a new direction. Nebula are inconceivably large clouds of gas, deep in space and under intense heat and tremendous pressure they condense, flatten out and infant stars are formed in the middle and planets are formed in the outer layers. What is left is a young solar system with the star in the middle and its orbiting planets making up the outer circle. The planets that were closest had rocky surfaces and the ones farther away were made of gas. At least that is the model that seems to be commonly accepted. It is also the accepted theory that explains the composition of our solar system. The planets closest to the Sun, like Mercury, Venus, Earth and Mars retained their rocky surfaces because they when they were young they were all too hot for elements like water and methane to collect in significant amounts. When the smaller planets became older and cooled they could not retain the large amounts of hydrogen which is characteristic of some of the larger planets like Jupiter and Saturn or so the theory goes.

With the passage of time and the discovery of more than a thousand new solar systems, the commonly accepted theory of how the solar system was formed is now being challenged. After the close observation of many new solar systems astronomers found that some of the larger gas planets such as Jupiter are closer to its host star and some of the smaller rockier planets like Mercury are farther away from that star, which is the opposite of the way our solar system was formed.

In fact, some of those planets in distant galaxies are larger than Jupiter but are closer to their host stars than Mercury is to the Sun. Seems the more we know the more we don't know. Being so close to a star and larger than Jupiter makes these big distance planets violent places, winds are super-heated and blow at tremendous speeds. Temperatures can exceed 1800 degrees Fahrenheit and the winds howl and more than a thousand miles an hour, it is an environment said to be so hostile that it can rain glass. On Earth temperatures of around 3,200 degrees Fahrenheit are required to form glass.

These big distant planets could have been drawn to their stars before those stars were even finished forming. The planets that are drawn to these stars collided with other planets and in the process formed even larger planets. This theory provides a new model from which astronomers theorize how solar systems form. In short it seems many new solar systems are formed in a reverse fashion when compared to ours, with large gaseous planets nearest theirs stars and the smaller rocky ones farther out.

Shotguns in Space!

Why would a country send its citizens into space armed with guns? The answer would lie with the Russians. The cosmonauts were given hand guns that were actually sawed off shot guns for protection against bears. No, the Russians didn't anticipate encountering bears in space, instead the powerful hand guns were given to the cosmonauts to use when they returned home. They Russians were issued what is called at TP-82 survival pistol which was really a sawed off shot gun made into a hand cannon. It was completely capable of taking down a large bear or a group of hungry wolves, which apparently were one of the many risks associated with the mission. Often times the Russian space crafts would not land where they had intended to touch down, but instead came to rest in the far out regions of the Urals or other remote areas of Siberia. Why the Russians didn't point their returning crafts into Earth's regions like its oceans that can cushion returning capsules is not known.

In 1965 two cosmonauts were stranded for days in the heavily wooded and extremely cold region of the Urals. They landed 600 miles off target in an area inhabited by grizzly bears and wolf packs. All the two cosmonauts had was one 9mm pistol for protection which apparently did not inspire the confidence that a TP-82 would have. So the cosmonauts came to terms with the fact that they were lost out in the wild back country of their home land. Fortunately, the feared grizzly bear or wolf pack attacks never materialized. In retrospect it seemed the cosmonauts' problems started well before they returned to Earth.

During the flight one the cosmonauts just completed the first human space walk and in the process nearly killed himself. Seems he had a problem with his space suit that malfunctioned and inflated to such a large size that it was life threatening. The suit was so blown up that the cosmonaut could no longer fit through the door way inside the capsule. So he took it upon himself to deflate the suit without permission from mission control and again nearly killed himself after contracting a case of the Bends. The Bends is condition that can be fatal and is brought about by rapid decompression. The pressure inside of his suit rapidly fell so low that it equaled what little pressure is on top of Mt. Everest which is a place called the "death zone." But rapid decompression was not the only problem they had to deal with, there was also the concern of a lack of oxygen, either one by itself was sufficient to cause death never mind a combination. Fortunately, no one died in space and their biggest peril was still ahead when they touched back down on Earth.

Their space craft malfunctioned during the descent but the crew managed to keep the craft under control and survived a very hard landing. When the door opened, sub-freezing cold air rushed in and to make matters worse their heater was not working. Imagine that, a broken heater on a malfunctioning space craft. Both cosmonauts looked out into the frozen wilderness, both were familiar with that area and both knew it was mating season for the local animals that might try to eat them. The 9mm pistol that had been issued to them would not have provided an adequate defense against a several hundred- pound bear, extra aggressive because of the mating season. A pack of wolves was spotted about a mile or two away from the downed cosmonauts by the rescue helicopter. It was three hours before a rescue helicopter could reach the men. But they still were not out of the proverbial woods just yet because the forest was just too thick and could not be penetrated for an air lift. Three long, cold days went by before the two men finally skied to a location where they could be airlifted to safety.

Come to find out there was more to the Soviet gun story than initially met the eye. For years its cosmonauts had carried a special survival weapon into space. The classic weapon had a folding stock that doubled as a shovel and an attachment that swung out as a machete. It also had three barrels. It would not be used if everything on the mission went according to plan but that did not always happen. When things went smoothly the gun was usually presented to commander of the mission as a gift. According to a number of reports, guns were never carried on American space flights which might be hard to believe given the popularity of guns in America. Instead the Americans sometimes carried a machete in case of a jungle landing.

The Russians were well trained in the use of their survival hand guns and years after the Cold War ended Americans were allowed to take part in some of that training. One of those reports indicated that the Russians seemed to have an endless supply of empty vodka bottles to use as target practice. And with the commission of the International Space Station came the sobering question: if guns should be carried in space regardless of the reason?

Presently there are guns on the International Space Station was well as the Russian craft that transports astronauts to that station. The serious question becomes apparent when one space flight in particular is recalled. There was a payload specialist who became severely depressed and then suicidal over an equipment malfunction. At one point the payload specialist tried to open one of the doors on the space craft which could have killed everyone inside. The payload specialist had to be wrestled into submission by another crew member. The incident happened years ago but the question remains, what if that payload specialist had a gun?

Out of this World Insurance!

Who takes out an insurance policy for damages or injuries caused by supersonic objects falling from the sky? Russia. Russia took out the policy on its Mir space station after it was decommissioned. It was 2001 and the space station had been in orbit for 15 years. Mir had been the pride of the Russian space program for most of that time. But after a decade and a half of cosmic wear and tear Mir had deteriorated. So part of the plan to decommission it meant Mir would be brought back down somewhere over the south Pacific, at least that was the plan. But the Russians also had a backup plan in the form of a 200-million-dollar insurance policy. Some wondered if that was going to be enough considering the amount of property damage and personal injury that could have been caused.

Here's some of the facts underwriters considered before the policy went into effect. Mir weighed about 150 tons, it would travel back to Earth at more than 10,000 miles an hour. If it hit something or someone the results would likely be catastrophic. According to the Russians, the chances of some sort of mishap were around 1 in 5,000. The encouraging possibility was Mir was expected to burn up in the atmosphere and what debris survived reentry was expected to land in the Pacific Ocean.

Still, more than a thousand fragments with a collective weight of 27 tons had a chance of reaching Earth's surface. Japanese officials didn't like those odds and mildly protested, representatives from several countries voiced identical concerns. Problem for Japan was Mir was expected to fly over its country during its final pass in low orbit. So the Russians took out a 200-million-dollar insurance policy that was underwritten by three of its nationally known insurance carriers. The premium on such a policy was said to be 1.4 million dollars. In fact, the 200-million-dollar policy was more than the entire budget for the Russian space program which at the time was 48 million dollars.

There were a wide variety of estimates when it came to taking a closer look at the chances that Mir might cause damage or injury. Some experts say you have a better chance of winning the lottery than you would getting struck by a piece of space debris. The satellite industry works with something called ground hazard which refers to the chances of someone on Earth getting hit by orbital debris. The ground hazard calculation method seems to be the industry standard which sets those chances at 1 in 10,000. Those are the same odds as finding a four leaf clover on the first try, not bad when you consider the odds from dying from heart disease for Americans are 1 in 3. So were the Russians playing the odds? The only difference was lives and property were on the line instead of money. Countries like Japan and Australia were hoping the odds were their favors.

One Russian official estimated the chances of someone on Earth getting struck by orbital debris were much less than getting hit by something that fell off of an airplane. Chances of debris from Mir hitting a person on Earth were 1 in 5,000 compared with the 1 in 10,000,00 chance of getting struck by something that fell off an airplane that flew overhead. That comparison did not inspire confidence and neither did an examination of the record. In 1978 debris from a Russian satellite crashed into the Canadian Arctic. One year later debris from an American satellite was scattered over parts of Australia and in 1991 Mir's predecessor slammed into the Andes mountains. Seems the Japanese had good reason to be worried but there was not a lot that could be done.

So Japan's foreign minister asked the Russians to provide detailed information regarding the decommissioning of Mir. After being armed with the requested information Japan stated that it would take all due precautionary measures, whatever that meant. Realistically how can you protect yourself from objects that fall from the sky? You just can't put on a hard hat. In the spring of 2001 Mir came back down to Earth somewhere over the South Pacific and left behind no personal injury nor property damage, just a trail of debris in the ocean that was described as a swath of buck shot.

But who would have been liable if the Sky Lab research station caused injury or damage when it was strewn over parts of the Australian Outback when it was decommission in 1979? The United States would have been obligated to pay under the 1972 Convention on International Liability for Damage Caused by Space Objects, just as the Russians would have been liable for any problems when Mir came back down.

Problem is only one nation had ever been sued after one of its satellites fell back to Earth and hurt someone so there is no real legal precedent there in part because the guilty country never paid the complete fine. In 1997 a woman claimed she was struck by an asteroid, which was later determined to be part of a previously launched rocket but compared it to a tap on the shoulder, so there was no pain or suffering to generate compensation. But the old saying "better safe than sorry" is the main reason why nations secure insurance policies for their satellites.

Here's the main gist of what happened during Sky Lab's reentry. It burned up over the Indian Ocean and parts of Australia. Some large chunks survived reentry and landed on the ground. No one was hurt. Some of that debris landed in a small town in Australia and NASA was charged a 400 dollar fine for littering. NASA never paid, maybe because of the same budget constraints that caused it to erase the tapes of its moon landing! But the fine was paid years later by a

DJ in California who collected money from his listeners to pay the Australian town what it was owed by NASA.

In 1978 a Soviet spy satellite that was designed to track US submarines came down. There was a malfunction on the Soviet satellite that caused serious concern from several governments across the globe, in part because there was a nuclear reactor on board. The satellite crashed over sections of Canada and contaminated some of that country with radioactive debris. There were no reports of personal injury nor property damage but the Canadian government billed the Russians 6 million dollars for the cleanup. The Russian's eventually paid half of the fine.

Debris from a NASA satellite came down in 1979 but crashed into the Atlantic Ocean. No one was injured and no property was damaged. In 1991 a Russian satellite that weighed 22 tons came back down after it broke up over Argentina. Debris was scattered over a small local town but no personal injury nor property damage were reported. In 2003 the Space Shuttle Columbia exploded shortly after takeoff and scattered debris over parts of north Texas. Everyone on board was killed and the disaster signaled the second fatal accident in the 30-year history of the shuttle program. No one on the ground was injured but debris was scattered over several states and the Gulf of Mexico.

In the fall of 2011, a satellite designed by NASA to research the upper atmosphere came back down. It weighed about 6.5 tons and more than a thousand of its pounds survived reentry. At that time officials said the odds of those pieces hitting someone on the ground was about 1 in 3,200. Once again space exploring nations play a game of chance with people on the ground and once again people on the ground beat those remote odds. But as any odds maker will tell you, sooner or later the law of averages will catch up.

If You See It Coming, Run!

But what about those rare cases when people are actually hit by objects that fell from space? Lottie Williams can give you a first hand account. The rare occurrence happened on a winter's night 1997. She and two friends walked through a park around three in the morning and looked up just in time to see a fire ball streak across the horizon. They were in Tulsa, Oklahoma and they thought they just saw a shooting star. They stood in awe and were taken by its beauty. But a short while later their moods changed. They were still walking through the park when Lottie Williams felt something tap her on the shoulder. Her first instinct was to run, she thought she was targeted by some sort of attacker lurking in the dark. Then she heard an object hit the ground behind her but she saw no one. Little did she know that what she thought was a shooting star minutes ago was actually a US satellite reentering the atmosphere and the sound heard in back of her was part of the craft that just crashed into Earth.

She had just beaten the 1 in one trillion odds of getting hit by orbital debris, that according to one expert who calculates those kinds of chances. Apparently the risk would be lower, like 1 in 10,000 if the space debris were bigger than what landed. According to NASA there are some 18,000 objects in orbit that are larger than a softball and 300,000 that are the size of a nickel and in total there might be more than a million pieces up there. Seems we all are taking our chances, no matter how small the odds.

Ms. Williams said the object from space was about the size and weight of an empty soda can, which put it somewhere between the size a softball and a nickel. She said it looked like fabric but sounded like metal after it landed. She did her research, got in touch with the experts and found out that what had lightly impacted her shoulder was actually part of the Delta II rocket. It had been launched the year before and the piece that came down was actually part of the rocket's material that was used to insulate the fuel tanks.

Apparently Ms. Williams was no worse the for wear, she was not injured and has an amazing one of a kind story to tell. She is the only person struck by orbital debris. She also has the ultimate conversation piece: that orbital debris that struck her to confirm her story. Her advice to others who might be in her situation: "if you see it coming, run."

Hot Hand!

More than 10 years later and half way around the world comes another story of objects falling from space and hitting someone on the ground. This time it was a 14-year-old boy in Germany who was struck by a meteorite. The boy was on his way home from school when he saw bright ball of light in the sky that seemed to be headed straight for him, which it was. The object actually hit the boy in the hand. It was red hot and about the size of a small marble that measured about half the size of his fingernail.

It then ricocheted off of his hand and embedded itself in the ground. Even though the object was less than inch wide it left a crater a foot wide in the ground. It was reported that the small object was traveling at 30,000 miles an hour before impact, chances are the meteorite slowed down considerably before it hit the boy. The boy did not lose his hand but was left with a nasty scar about three inches long where the meteorite hit him, chances of getting hit by a meteorite are calculated to be 1 in a million.

So what was it like? The boy told reporters after he saw the bright light he felt intense pain in his hand and then heard a bang like the crash of thunder. Meteorites often travel several thousand times faster than the speed of sound; the speed of sound is considerably slower at 768 miles an hour. So the boy first felt the impact and probably thought what in the world was that and then heard the sonic boom sometime later, most likely still not completely understanding what just happened. Judging from the 3-inch mark, the meteorite grazed the hand and didn't hit it directly or else the injury would surely have been worse. That meteorite was probably 4.5 billion years old and dates back to the formation of the solar system. It was also likely a lot bigger before impact because most of it probably burned up during reentry.

The boy said the impact sent him flying and the sound made his ears ring. The small meteorite was found buried in the road in the town of Essen, Germany. Scientist retrieved the space rock and confirmed that it was a meteorite. They said most of these objects don't make it all the way down because they usually break apart in the atmosphere and those that do land in water 6 out of 7 times. According to the experts the meteorite probably slowed down to terminal velocity, which is the speed an object will free fall through the atmosphere without any aerodynamic assistance, estimated to be around 128 miles an hour. They say there's not much chance the boy could have lived through an impact much faster than that, let alone speeds of 30,000 miles an hour, which is almost a dozen times faster than a bullet fire from a rifle. At that speed even a glancing blow by a pee sized object would leave a hole about the size of a basketball. There is also no concrete estimate given regarding how much time went by from when the boy first felt the pain from the meteorite strike to when he actually heard the meteorite's sonic boom.

Researchers say here's what probably happened. The meteorite hit the ground first and then a fragment hit the boy's hand. There didn't seem to be any independently corroborated reports from witnesses, which are not needed to verify the case but either way the story has been recorded as a documented fact and researchers say there is no reason not to believe the boy.

Rude Awakening!

The only other known report of a person being struck by a meteorite was when a woman in Alabama had a rude awakening. The year was 1954 and she has the dubious title of being the first person known to have been hit by a meteorite. There is a chance prehistoric humans were hit by meteorites but there is no record of that because it was prehistoric. Apparently a dog in Egypt was killed by one in 1911.

Elizabeth Ann Hodges has a tale to tell. As she lay asleep one night a meteorite crashed through her roof and into her house, bounced off of a few things including a radio before it landed on her. Mrs. Hodges was asleep on the couch. It is amazing she was not killed considering the size of the meteorite and the speed at which it traveled. The space rock was measured at 8.5 pounds and seven inches in length and they can break through Earth's atmosphere at more than 20 thousand miles an hour. But this one was slowed significantly by the atmosphere and then her house before it landed on Mrs. Hodges. She was hurt but did not suffer permanent injury, she had a bad bruise on her leg and hip, not to mention a gaping hole in her living room ceiling.

Besides being the first documented case of a meteorite striking a human this was also the most widely covered. The story made international news. Neighbors in the area reported seeing a bright flash of light before impact. Some described it as a Roman Candle streaking across the sky and trailing smoke, others saw it like a giant welding arc. Other people thought it might have been a plane crash and suspected the Soviets. At that time Cold Way hysteria was at an all-time high so authorities confiscated the meteorite and turned it over to the Air Force.

A geologist who worked at a near-by mine was called into to investigate and he confirmed that it was a meteorite. By the time Mr. Hodges got home from work there was so many curious on lookers that he had to push them off of the porch just to get inside. A short while later Mrs. Hodges was hospitalized, apparently due to stress that the incident caused and the subsequent public reaction and media coverage. She died years later of kidney failure. Her husband suspected her ill condition was brought about by the traumatic incident of the meteorite impact and later told a reporter that she never recovered from the event.

There are at least two other stories within the main story here. After the meteorite strike came the legal battle over who owned the space rock. Mrs. Hodges rented the house so legally the owner of the home owned the meteorite. A law suit ensured, time passed, public opinion sided with Mrs. Hodges and the home owner eventually gave up the effort, so Mrs. Hodges was able to keep the meteorite. But by that time public interest in the story had dropped off and it seemed nobody wanted the meteorite any longer. So Mrs. Hodges used the meteorite as a door stop. One can only imagine her disappointment, at one point she probably hoped the meteorite was going to be her ticket to a better life only to find out that her "ticket to ride" was now a door stop. In fact, a smaller meteorite fell on a neighbor's property and the neighbor sold the meteorite and was able to buy a new house along with a new car. And the neighbor's meteorite was about half the size of the one that made Mrs. Hodges famous. Convinced that her pay check from space was gone she donated the meteorite to a local library where it is on display to this day.

The only person that seemed to profit from the meteorite impact was that neighbor who found the smaller asteroid. Here's the story of Julius McKinney. He was a Black man in 1950's Alabama who faced all of the problems a Black person would face at that time in the south. He was on his way home from work one night in a mule drawn wagon when the mules motioned at what appeared to be a black rock in the middle of the road. Mr. McKinney got off the wagon and kicked the rock to the side, got back on the wagon and went home without thinking much about it. When he was home he heard reports about the meteorite that struck his neighbor so he immediately went back to the site, found the rock and took it home and let his kids played with it.

Mr. McKinney was a farmer, and as a Black man in the American south during the 1950's he didn't trust many people. Besides family members he apparently only trusted his mailman so he told him about the story. The mailman delivered like never before, arranged an attorney to represent Mr. McKinney and to negotiate the sale of the meteorite. Mr. McKinney then went public with what he discovered in the middle of the road that night.

The meteorite was purchased by a representative of the Smithsonian Institute in Washington, D.C. Mr. McKinney, who probably had enough of wagons drawn by mules bought a car, which might have been his first one. He also bought a new house, no doubt everyone in the family enjoyed it. The movie rights to the story were sold but none was ever made. Locals called the incident the biggest thing to literally ever hit that small town in Alabama.

Consolation Prize!

There are stories that qualify for consolation because they involve meteorite strikes but no one was hit or injured. In 1982 a husband and wife in Connecticut heard a loud noise that came from the front part of their house. When they went to investigate they had a hard time believing what they saw. The house was a mess, a hole was in the ceiling, plaster, dust and debris were all over the room. They called the fire department which came to investigate and found the problem under the dinner table, a 5-inch meteorite. Seems the little space rock hit with such force that it ricocheted all over the home and even bounced through the ceiling twice, hitting the attic upstairs before coming to rest downstairs under the dinner table. Sounds like it was a ball in a pin ball machine. The impact was so explosive that six smaller fragments from the meteorite landed inside the couple's vacuum cleaner, which was probably going to be used to clean up the mess.

In 1992 a fireball was seen burning up in the sky over New York before it crashed into a parked car. The woman who owned the car was inside the house and ran outside after hearing the loud boom. She too couldn't believe what she saw. There was a sizable hole in the rear end of her Chevy Malibu. She took a closer look and there was another sizable hole in the ground right under the car where the meteorite burrowed into the Earth.

The meteorite was shaped like a football, was about the size of a bowling ball and weighed some 28 pounds. Judging from the size of the hole in the Chevy Malibu it sliced through the metal like a hot knife through butter and kept going right into rocky ground below. Apparently it smelled like rotten eggs. Witnesses said the bright flash crackled like a large sparkler as it sped overhead. Scientists took a look, confirmed it was authentic and said it likely came from the Asteroid Belt, between Jupiter and Mars, some 365 million miles away.

Somehow there was money to be made off of this meteorite impact. Granted meteorites are fairly common but a car hit by one is not. The18 year old woman sold her Chevy Malibu for 10 thousand dollars which was probably more than what it cost new. So the moral of the story could be your car will certainly depreciate unless it gets hit by a meteorite.

In 2003 a meteorite bigger than a refrigerator exploded in the sky over Chicago and showered parts of the city with bits and pieces of its fragments. It happened about 30 miles south of downtown and it spawned all sorts of stories from all sorts of witnesses. One teenager woke up after a five-pound meteorite crashed into his bedroom and stopped just a few feet from where he was sleeping. There were other reports where meteorites broke windows and struck cars, police departments fielded numerous calls from people witnessing the ongoing explosions. The light show was so large it was seen in Ohio, Wisconsin and Indiana. Pieces of the meteorites were scattered in an area some 80 miles long and 20 miles wide. Some of the pieces that made it all the way down were 7.5 pounds. There were no reports of anyone seriously injured. Apparently everyone on Chicago's south side beat the odds that night.

The Big Rip!

The Big Rip theory is one of the latest models to challenge the Big Bang theory of how the universe might end. While the Big Bang suggests that the universe will collapse on itself after it had expanded to a great distance, the Big Rip theory suggest something to the contrary where the universe expands to the point where it rips, hence the theory's name.

Proponents of the Big Bang theory say the universe will keep expanding and the amount of its matter will determine how fast it will grow. The more matter might mean a slower expansion in which the universe could collapse on itself to where it reaches one small single point and then explode into another Big Bang.

Not so with the Big Rip theory which could be compared to a runaway vehicle with no breaks, the vehicle is not going to stop on its own. Under this scenario the universe just keeps growing and expanding until something gives and the more it expands means the faster it expands. When something does give, a galaxy or a solar system would be ripped apart. It is thought that the ripping apart process would feed upon itself, first galaxies are torn apart and then solar systems, followed by planets and eventually even atoms. The Big Rip theory is incomplete but represents one of the most alarming ideas on how things will end. The process of this compete ripping would leave the universe without structure, no galaxies, no planets not even atoms.

Much of the Big Rip theory depends on a type of dark energy called phantom energy. It is a hypothetical form of dark energy that possesses negative kinetic energy. Kinetic energy is something that is produced by everything that moves or put another way motion creates kinetic energy, so negative kinetic energy might be just the opposite and is created when something does not move. Phantom energy is when the surrounding pressure is greater than the surrounding density. It is this pressure that allows the universe to keep expanding until it rips apart. That explanation is barely sufficient and threatens to over simplify a very complex process but this is the limitation of our current understanding.

Problem is we can't see dark energy, just like we can't see phantom energy and we don't know the ratio between the pressure dark energy generates and the density of that energy. So how do we know if it evens exists? We don't, because it is a theory, albeit a theory founded on scientific models and observations but a theory none the less. The theory will not be proven or disproved anytime soon because the Big Rip could take place some 22 billion years from now. By then Earth might have been engulfed by our Sun, but let's imagine that our planet is still around at that time, Earth would be ripped apart shortly before the solar system was destroyed. So what happens after the Big Rip? One school of thought is that the matter left behind, both big and small will start to recollect itself under the force of gravity and after enough time has passed there might be the recreation of objects in space similar to the ones that were already there.

To understand more about the Big Rip, we need to know more about dark energy. Again, problem is we don't know if dark energy even exists. And if it does we still don't know what dark energy is made of or how it works. Scientists only know what dark matter does not have, which are some of the basic elements contained in normal matter, like neutrons and protons. Despite the gaping holes in our existing knowledge, scientists think that dark energy makes up about three fourths of the universe. It was not until the 1920's that we figured out that the universe is expanding, so needless to say humanity's collective understanding of the universe is still very much a work in progress. But we do know that gravity is one of the universal laws and that gravity should slow the expansion of the universe.

So much of this is still a mystery but scientists theorize that dark energy makes the universe expand and dark matter accounts for the material inside the expanding universe. Scientists also don't know the source of either dark energy or dark matter but are relatively certain these two forces play a significant role in the way the universe is shaped and behaves.

Dark energy is thought to be a repelling force that counter acts gravity and causes the universe to expand at an ever increasing rate. Dark matter is thought to be the material that holds stars and galaxies in place, it is matter that does not produce light nor interact with light and is thought to exist in space in one of several forms. Dark matter is not made up of particles like protons, electrons and neutrons that comprise normal matter. Normal matter is thought to contain the building blocks of life, including atoms and their smaller elements and everything that can be seen in the universe is made up of this material. Problem is most of the material in the universe cannot be seen. So the mystery of dark matter remains just that.

Scientists also theorize that dark energy is pushing the farthest galaxies even farther with the rate of expansion accelerating. So the farther out the universe is, the faster it travels, its speed continues to build until all of its matter gets torn apart which includes, solar systems, stars, planets all the way down to sub atomic particles.

Some scientists say the universe will be torn into tears by the Big Rip, the tears being what's left of matter like planets and stars. Galaxies would be flung out into the abyss at velocities approaching light speed. Eventually these galaxies are so far out there they can no longer been seen. Again all this is thought to be looming billions of years in the future, compared to our own doomsday on Earth set to be 7 million years away when our Sun expands and engulfs the planet.

Cosmic Conspiracy Theory!

All the talk about dark matter and dark energy once again brings up the question is there some natural force that keeps the secrets of the universe hidden from science. Think about dark matter, which is something we cannot see but is thought to make up most of the universe. Think about dark energy, again something we cannot see but is thought to guide, shape and expand the universe. Think about the Higgs boson particle, the sub atomic particle thought to give mass to all matter, again something we cannot see but is thought to be fundamental to life. Black holes thought to be fundamental to universal function but yet we also cannot see them. So we can't see any of these essential things and we are not even sure if they exist. How can you possibly hope to learn about such matters under such bleak circumstances?

Think about the great distances of space's inhospitable environment that are impossible to traverse. Get the picture, is the universe sending us a message? There is a good reason why we know so little about the bottom of the ocean, because its extreme environment prevents us from learning any more. Is that not an applicable comparison to space? If the answer is yes, it would be quite the blow to our ambitions of space exploration. Remember the space race that lead to the first moon walk! At that point many scientists were certain that was the start of big things to come, like visiting other planets and starting human colonies out there. Well 40 years after the fact we are no closer to realizing those dreams than when they first presented themselves.

Space is too formidable and technology is too feeble and the limitations of the human body are too many to get too excited about space travel. Some of the great thinkers on the subject say it'll be at least one hundred years before humans could colonize other planets. That would be nice but it could be a lot longer than that. Humans have an extremely slow rate of evolution. Simply put our bodies are not adapted to space and there's no indication that they ever will be. Even though technology has made great advances since we've landed on the moon, our current understanding of such matters is too inadequate to travel much past the moon.

So we could be in a race against time, that according to some of the experts who say humans will make Earth uninhabitable due to nuclear war, climate change, genetically engineered viruses and the pollution of the delicate environment just to name a few. They say this could happen within 500 years and if we are to survive we must get off the planet. Problem is we can't because we don't know how. Those same experts say in light of this grim reality we better start taking better care of Earth which is the only home we have. They say we should stop chasing the cosmic dream of trying to find a new home somewhere out in the galaxy and instead reallocate funds to support less ambitious and more realistic space programs.

Think about all of the billions of dollars that has been spent on space exploration, most of it is tax payer money and the average tax payer probably cannot see the value in that kind of spending because they're just trying to make ends meet. The marvels and wonders of space don't carry a lot of weight with the homeless or the jobless or the terminally ill or even with the working poor. And what about people in the third world countries devoid of the basics, just trying to stay alive by finding enough food to eat and clean water to drink with their numbers estimated at 800 million. It is hard to justify spending trillions of dollars to send a hand full of people into space while millions die of starvation. Again is space telling us something?

One of the Biggest Small Monuments!

There is a small monument in space that carries a big message from humanity. It is on the moon and was placed there in the summer of 1971 by an American astronaut. He was part of the Apollo 15 mission and he and others wanted to pay tribute to the fallen comrades, those who died in space, regardless of nationality so that our knowledge of the universe could continue to grow. It is called the Fallen Astronaut monument. It is an aluminum sculpture a little more than 3 inches in size and it depicts an astronaut in a space suit. Next to the monument is a small plaque that lists the names of the 14 people known at that time to have died in space. The number has since risen to more than 20.

The crew kept everything a secret until their mission was a success. When the public was notified a well-known museum wanted and received a replica of the sculpture. The artist that designed the sculpture wanted to sell 950 copies of the statue, presumably for personal gain, although later in a written statement he denied that he was motivated by money. NASA protested the artist's effort due to its policy that prohibits commercial advantages gained from its space program. The artist finally agreed with the prohibition. Turns out that years later the financial ambitions of the artist belied his true intent. Apparently out of the 950 replicas in his possession the artist intended to sell them for $750 dollars apiece. He also had plans to sell second editions of the replicas much cheaper through catalogs. But again NASA put the k-bosh on his get rich quick plan.

Here's how it all began. One of the Apollo astronauts met the artist at a dinner party where the artist agreed to create the statute. As is often the case with verbal commitments the two men disagreed over what was agreed upon. The astronaut said they had agreed to make a small, lightweight statute that could withstand the extreme temperatures of space. It would include the names of the 14 men, the Americans and Russians who died in space. The statute would not identify a male or female or any ethnic group and was to be as generic as possible. The statue would also be placed laying down. Everything was to be a secret so as to avoid NASA's policy prohibiting the commercialization of its efforts. The statute was to be smuggled on board the space craft.

The artist remembers the details differently. He thought the statute was to represent all of humanity and not just 14 astronauts. He also thought the stature would remain upright and not laid down and that it would be commercialized. More than 40 years later the two men were still in disagreement. Both men presumably knew about something called a contract but didn't sign one which could have cleared up this kind of problem.

The statue and plaque should stand the test of time and could be on the moon for eons, maybe a million years or longer. Reason being, there is no air, wind, rain, erosion, atmosphere or volcanic activity to change anything on the lunar surface. Again the artist wanted his sculpture standing on the moon instead of laying down so that it could better represent all of humanity, can only hope his disappointment won't last as long as his statue.

Where there is One there's usually More!

Turns out we have left a lot of stuff on the moon which raises the question are we trashing the lunar surface? Read and judge for yourself with this incomplete inventory. More than 70 space craft, rovers, modules and such, 5 American flags, 2 golf balls 12 pairs of boots, TV cameras, film magazines, 96 bags of urine, vomit and feces, numerous cameras, improvised javelins, numerous hammers, backpacks, insulating blankets, utility towels, used wipes, personal hygiene kits and empty bags of food make up what has been left behind on the moon.

When Apollo 12 went to the moon its crew took with them a number of drawings including one done by Andy Warhol. Warhol's obscene drawing is a crude representation of the male genitalia that looks like it was drawn by a third grader. Mr. Warhol was an American artist known for his work in the pop and visual genera of art. He was supposed to leave behind his initials or a special version of a space ship but instead he left behind a crude drawing of his penis. NASA did know it until it was too late because the pictures were smuggled on the space craft.

In 1972, Apollo 17 Astronaut Charles Duke took a photograph with him to the moon. The picture was of his family. But he must have forgotten how tough it is on the moon. The sunlight, the temperature and radiation are so intense that the picture quickly faded. So what was left if is a blank portrait of the Duke family. The picture is probably bleached completely white by now due to the intense unfiltered radiation from the sun. In the summer of 1971 a piece of lava rock from a lake in Bend, Oregon ended up on the moon. Astronauts used to train for missions to the moon by collecting rock samples in Bend, Oregon and in this case a building inspector was part of the program. When the astronaut was dispatched to the moon the building inspector with whom he trained asked him to take that rock from Oregon with him as a favor.

The Apollo 11 crew brought messages from 73 world leaders including the Queen of England who like the other heads of states all wished the astronauts well. It is a one of a kind goodwill message contained in a disc about the size of a 50 cent piece with a worth that is priceless.

Is it X or Nine?

Science just like religion sometimes asks you to takes leaps of faith in the absence of evidence and that seems to be the case with something called Planet Nine, also called Planet X but some astronomers say the two are not the same while others disagree which in no way provides clarity but please be patient because transparency will come. Nine or X is a newly discovered planet but no one has ever seen it. So how was it discovered? Apparently from the way it influences other visible matter, like planets and asteroids that happen to be the general vicinity. This method is also used to detect black holes, which also cannot be seen.

One possibility why we can't see it is because the new planet is a long way away, thought to be billions of miles from the sun. Its orbit is so far out that it might take it 20,000 years to circle the sun just one time. So the last time this planet might have come close enough to see from Earth no one cared because they were trying not to freeze to death during the last Ice Age. So a year on Planet Nine could actually take 20,000 years on Earth. Its orbit is so vast that is thought to be 20 times larger than Neptune's. Neptune is the planet farthest away from the Sun and has the largest orbit of any planet in the solar system estimated to be some 2.8 billion miles from the Sun.

But again it might not exist because it has not been seen directly. Although the law of physics suggests that it is out there and here's why. There has been strange movement in the Kuiper Belt, the area on the outer perimeter of the solar system. There are two places where asteroids collect in the solar system, the Asteroid Belt, located between Jupiter and Mars and the Kuiper Belt located just beyond Neptune. The Kuiper Belt is the larger of the two, thought to be 20 times as wide and 200 times as massive than the Asteroid Belt. The Kuiper Belt would be closer to Planet Nine and therefore more susceptible to its gravitational influence. The Kuiper Belt is relatively stable in part due to the gravity of the sun. But some of the asteroids in the belt behave as if they are not influenced by the sun's gravity but instead as if they were orbiting some other large body in space. That other large body just might be Planet Nine. Scientists tried to answer questions but they only raised more questions. For example, scientists say that there is a very good chance that Planet Nine might not be a planet but don't know what else it could be.

Planet Nine would be the ninth planet in the solar system, that after the previous number nine was Pluto, demoted to dwarf planet status because it didn't meet all of the requirements. So for the first time in 150 years the inventory of the solar system is incomplete. It was thought that there were no planets beyond Neptune and now there might be one 10 times the mass of Earth.

Then there is the theory of Planet X, which again raises the question could Planet X really be Planet Nine? The Planet X theory has been around longer and posits that our solar system contains a large undiscovered planet. But according to theory, Planet X's orbit will take it very close to Earth. That does not seem likely with Planet Nine because its orbit is too far away. Theorists also claim that at some point Planet X will collide with Earth which is known as the Nibiru Cataclysm. There is nothing to suggest that Planet Nine will collide with Earth. Probably the death nail for the Planet X theory is that it will collide with Earth in 2012 but to painfully over state the obvious 2012 has come and gone.

Planet Nine might have an extremely elongated orbit which could explain why it has not been seen by astronomers. Simply put, the elongated orbit which covers great distances, might put Planet Nine so far away that it just can't been seen. Since it might take 20,000 years to complete one of its orbits, it could have come close enough to see from Earth 20,000 years ago but at that point cavemen didn't have telescopes. So in theory its next closest approach could be our first opportunity to confirm its existence.

So if Planet Nine exists, where did it come from and how did it end up way out there? According to one theory there were four gas planets formed in the early solar system, Jupiter, Saturn, Uranus and Neptune. Those four planets grabbed all of the gas around them in order to take their present shapes. But there could have been a fifth giant gas planet in Planet Nine and Planet Nine could have got too close to Jupiter and Saturn and was ejected by their gravities out into its present orbit. So it seems the race is on, apparently the first person who sees it gets to name it and in the process would restore the original number of planets in our solar system back to nine made possible by the planet with the same name, Planet Nine.

Three on One!

A fire on board Apollo 1 turned out to be fatal and would be a precursor of the fiery deaths that would await future astronauts. Three astronauts were killed during that incident. The official cause of death was asphyxiation but as we mentioned earlier it sounds as if they were burned alive. Asphyxiation implies that the victims were overcome by smoke inhalation which often comes during a state of unconsciousness. The body of one of the victims was frozen in time, he was trying to unlock to the door to the capsule so it would be safe to assume that he was conscious when he was engulfed by flames.

All three men died in what NASA called a "non-hazardous" ground mission. Just like their Russian counterpart Vladimir Komarov, who was also killed in the same year, the Americans had well founded worries that their capsule was not safe. Their concerns were so serious that all three men, Ed White II, Gus Grissom and Roger Chaffee took a picture together before they were killed. They posed in front of a model of their craft and all three men were praying. Some critics of the disaster say the fire was bound to happen and NASA officials either knew about the dangers or should have known.

Pure oxygen was used in the cabin instead of a diluted mixture. The door hatch opened the wrong way which was inward instead of outward which could and did trap the astronauts inside. Flammable Velcro straps were also used. But at least one NASA officials said if the fire on Apollo 1 did not happen we never would have reached the moon, a statement apparently made to extract something positive out of something negative but few people really thought that statement carried any truth.

That fatal fire in January of 1967 resulted in several operational changes but it would be 18 months before NASA would send any more of its men into space. The fire flashed through the small capsule in an instant, the three men inside were immediately gone. They all knew that the craft was not quite ready but apparently perils like that went with the territory. All of NASA's astronauts had flight experience and many were test pilots. They were accustomed to seeing machines that were prototypes, or were under development or for whatever reason were just not quite ready for flight. It's not as if these men, American and Russian, had death wishes but possibly rather a commitment to duty that superseded the innate will for self-preservation.

On his last trip home one of the astronauts took a lemon off of a tree in his back yard. When his wife asked him what he was going to do with it, he said hang it on that space craft, presumably for good luck and when he returned to the launch pad that's exactly what he did. The morning of the test a foul odor was detected in the capsule which delayed operations for about an hour. Then there was a malfunction with the communications systems which prompted doubts from one astronaut who said something like how could they get to the moon if they can't talk effectively on the launch pad. The communication problem had not yet been fixed when the word "Fire" rang out. It was a flash, everything went up in an instant, bright flames took up the entire frame of one of the monitors, the crew struggled in vain to get out, help rushed to the scene but by the time the hatch was opened it was too late.

Obvious changes were made after the fire, pure oxygen was replaced with a nitrogen oxygen mix, flammable objects were removed from the capsule, the door was reconfigured to make it easier to open in a matter of seconds. Critics say these were common sense changes that should have been made before those three men were killed in that fire. NASA stated that what happened could have been avoided.

The Latest in a Series!

On a winter's morning in 2003 the Shuttle Columbia was coming home when disaster struck and killed everybody on board. The craft was flying upside down, a common practice in space that affords the crew a spectacular view of Earth as opposed to the other direction that showed only the pitch blackness of space. They were traveling at 17,641 miles an hour some 237,000 feet above Earth when a small hole appeared on its wing. Sensors indicated that the craft was overheating but it was too late. A stream of high velocity, super-heated upper atmosphere tore its way inside of the shuttle. The craft disintegrated and its seven astronauts were gone in an instant. It broke apart over Texas just minutes before it was to land in Florida. Debris was scattered over several southern states. Much of the world stood in shock, collectively thinking, not again, not another shuttle explosion that would kill everyone on board. It would be two years before another shuttle would fly again.

The investigation into the explosion determined that all of the crew, five men and two women were apparently doomed just minutes after takeoff. Those men and women had no idea that they were living on borrowed time and their parting views of Earth would be their last. A piece of foam insulation from the shuttle's fuel tank broke off and damaged the wing when it hit it. That happened on January 16, 2003. Two full weeks would go by without a problem and then on February 1, 2003 a problem appeared and there was no remedy. The disaster was every bit as bad as the one in January of 1986 when the Challenger broke apart just after takeoff and killed all seven people on board.

One of the changes that followed: Space Shuttles only flew missions to the International Space Station. If the rule had been implemented before the Columbia disaster its crew might have been saved, or at the very least they would have had a fighting chance. Flying to the International Space Station meant the crew had a safe haven in the case of a potentially fatal problem, like a wing damaged during launch, and at that safe haven they could either fix the problem or get rescued.

The same premonition that was present with the crew of Apollo 1 and with the crew of the Russian craft Soyuz 1, seemed to show its dreadful face with the crew of Columbia. The mission was plagued with problems well before launch. So many problems that the original mission was delayed for a full week. Did Columbia, just like Apollo 1 or the Soyuz 1 launch prematurely? Hind sight is always 20-20 but the answer seems to yes. Why did American and Russian space programs operate like they were in a race against time and feel pressure to commission craft that were not quite ready for the rigors of space? The Cold War had long since been over but yet that behavior persisted with something called the space race which was turning out to be more deadly.

Cameras on board the Columbia recorded the collision between the piece of foam and the wing. Similar collisions happened on three prior shuttle flights but those prior three flights landed safely. This time a few engineers thought the Columbia collision would end up being a catastrophe and they were right but unfortunately their concerns were ignored for two long weeks.

The first bits of Columbia fell out of the sky over Texas, a minute later came the crew's last communication and then seconds later it was gone, disintegrated in a fiery explosion that scattered debris over 2,000 locations in three different states. Their deaths were probably in the worst manners possible, exposed to the excruciating pain of fire.

An article in the on-line publication spacesafetymagazine.com recounted the incident with riveting detail. Columbia's crew apparently had no idea of what lay ahead just minutes away as they recorded a video of what would be their last moments as they reentered the atmosphere. Mission control also performed the last checks but less than ten minutes into reentry ground crews got clues that something was wrong. Temperatures in the hydraulic fuel of the left wing sky rocketed, followed by a loss of tire pressure on the left side meaning the Shuttle Columbia would not be able to land. Then all of the sudden communications went out, partial communication was expected but not total silence. Nothing seemed to work including radar, an indication that catastrophe was in progress. Ground control blurted out something lock down or close all doors which was part of protocol to control all information during emergencies. As that was happening the first reports started coming in of debris falling from space and racing across the sky which was the telltale sign of the disaster had already played out. A short while later came the confirmation in the form of a presidential address.

Come to find out after an investigation that the crew could have repaired the damage to the wing. The investigation also revealed a second option in that the crew could have been rescued. There was no reaction to the investigation results from NASA officials who for two long weeks ignored concerns from engineers about the damage that occurred during launch. All sorts of questions remain after the disaster and none will probably ever be answered including: Was it better for crew members to die unexpectedly or would it have been preferred to know what was coming to possibly say final good byes to loved ones on Earth through radio transmission?

Gould's What!

Chances are you've never heard of something called Gould's Belt. But if you look up into the night sky at the right time and in the right place you might be looking right at it. It's group of stars. The stars are in the constellations, Cepheus, Perseus, Orion and Canis Major. Gould's Belt is a big place, stretching some 3,000 light years across and it's made up of relatively young stars. The stars in this belt are about the same age as our sun and are about 1,500 light years away. One of the interesting things about Gould's Belt is that it holds the potential of increasing our galactic knowledge in general and increasing our knowledge about stars in particular.

Gould's Belt contains clouds called nebulae, as well as gas. One cloud is called a nebula, more than two are called nebulae. The belt appears to be on the lower edge of Orion's arm and is home to many of the brightest stars in the galaxy. Scientists think that nebula has been around for billions of years and even passed through the Milky Way. While in the Milky Way, gas was compressed inside the nebula, it became heated and pressurized, parts of the nebula started to glow and many stars were formed in the process. Stars are not usually formed alone but rather in groups or clusters and some in Gould's Belt are the brightest in the galaxy, in fact most of the big stars that are close to our sun are in Gould's Belt.

The stars form a partial ring deep in the Milky Way and many of those stars are 0 and B types. B and O stars are thought to be the brightest, biggest and hottest stars in the galaxy which explains why Gould's Belt is so luminous and so unique. B stars are blue, super bright and are up to 16 times larger than our sun. Their surface temperatures can be as high as 53,000 degrees Fahrenheit, which is about five times as hot as the sun's surface. O type stars are up to 90 times larger than the sun. The surface temperature of an O type star is as high as 89,540 Fahrenheit, almost nine times as hot as the sun's surface. And that ring of stars could be more like a disk of stars, reason being disks and spheres are two of the most common shapes in the universe and both are created by gravity.

The belt is thought to be 30 to 50 million years old and came from an unknown origin. One possibility is that it was formed when it collided with dark matter and a molecular cloud. Another possibility could be Gould's Belt was formed by two or more super novae, apparently one would not have been strong enough to do the job. A super nova happens at the end of a big star's life when it becomes really bright and explodes. The sun is too small to go supernova.

Gould's Belt was named after Benjamin Gould who many people say discovered it in 1879. But John Herschel was really probably the first. Mr. Herschel described a "zone of stars" and published his discovery in 1847. There seems to be no explanation why Gould got the credit and not Herschel but both astronomers made their observations in the southern hemisphere which made for the best viewing. But star gazers in the north should not feel neglected because Gould's Belt has a tilt to it that allows northerners a special view of the bright stars that they could not see if the belt were one a level plane. The tilt acts almost like a mirror that reflects the sun's light, and if you were to tilt the mirror at the right angle you'd get the full effect of the sun's blinding light shining right at you.

Our sun is inside of Gould's Belt but does not sit directly in the center. Stars that are much bigger than the sun are also in the belt and when they explode they can leave black holes and pulsars. Black Holes are areas in space where matter is compressed into a tiny space and gravity is strong that nothing can escape, even light. Pulsars are thought to be pulsating neutron stars that rotates so fast they can spin one thousand times per second. A neutron star is what is left over after a massive star collapses. Remember neutron stars are thought to be the densest and smallest stars in the universe. Many of the stars in Gould's Belt contain lithium which is an indication of their young age. Older stars burn hydrogen and helium and not lithium.

Gould's Belt circles the sun and is so luminous that it is compared to a brightly burning orientation beacon for the entire galaxy. One reason why Gould's Belt is so bright and so special is the type of stars it contains, most notably O and B types. O and B type stars are at the top of the seven tiered scale gauging how hot stars burn. Our sun is well down the scale second from the bottom in the fifth spot with a surface temperature of about 10,000 degrees Fahrenheit. They are part of a group called Main Sequence stars.

Main sequence stars are those that fuse hydrogen atoms into helium atoms and make up about 90 percent of the stars in the universe. They range in size from one tenth our sun to 200 times larger than the sun. These stars are not formed individually but are created in groups or clusters. How long a main sequence star burns depends on its size. The bigger the star, the more mass it has so the hotter it burns, which means the faster it burns through is fuel which also means it has a shorter life. The fuel is in the star's core which is extremely compressed and hot due to tremendous pressure and gravity.

The sun should burn for 10 billion years but a star five times as big may last only 10 million years. Again that's because the bigger star burns hotter and more quickly through its fuel. Some type O stars are so massive that they rapidly burn through their material in just few million years and almost all of them will go supernova when they die. O stars are also the hottest of stars and the rarest. About one in every 3 million stars is an O type star.

It is safe to say that most of the bright stars in the sky are in Gould's Belt and that includes those in Orion and Canis Major Constellations.

Orion and Canis Major are in constellations or groups of stars that form patterns which are recognizable. Gould's Belt traverses about twelve of these constellations which makes the belt a big, bright, important formation.

The next time Gould's Belt shows itself in the night sky we might look up and hope for another Earth like planet somewhere in the huge disk of stars. One would think that at some point the law of averages will reveal another such planet and what better place than Gould's Belt, arguably the brightest spot in the galaxy. Perhaps hidden behind one of its many big brilliant stars sits a lesser star, one like our Sun, and revolving around it might be a planet like Earth, with intelligent life that might look toward our planet and ask the same question we ask.

High Five!

There was a rare light show recently seen by millions but it was likely missed by many more. Most of the planets in the solar system were uniquely positioned in a line so they could all be seen at once with the naked eye. Five planets were on display at the time, called the high five. That doesn't happen very often. Mercury, Venus, Mars, Jupiter and Saturn were all on display in the winter of 2016. As an added bonus two stars were thrown in for good measure, which would be Antares and Spica. Those stars appeared as twinkling objects and lined up about mid-way through the line of five planets. A special appearance was made by the crescent moon. These cosmic shows don't get much more special than that, unless something catastrophic is about to happen like an asteroid impact.

The cosmic light show began around January 20, 2016 and ended about a month later in February. People had to rise early to see the show but the price of admission was right. Star gazers in the east rose around 5:30 am just before twilight to enjoy the sight which lasted for less than an hour and looked something like this. Looking to the eastern sky, the planet farthest to the left and lowest in the sky was Mercury, followed by Venus to its right and a little higher up was Mars all the way to Jupiter which was at the far right at the top of the picture and then Saturn. The distance from Mercury to Saturn is some 850 million miles but it was all there in plain view for star gazers to marvel at. The positions of the planets would slightly change but at one point you could see Jupiter rising first in the evening then Mars at midnight followed by Saturn, Venus and Mercury.

There was more to marvel at than just the sight. Remember all of those planets orbit the sun and their paths have been compared to the hands of a clock all working in synchronization. Mercury goes around once every three months. Venus goes around in seven months. It takes Mars two years, Jupiter 12 and Saturn brings up the rear at 30 years. Of course all of the planets are traveling at different speeds, Mercury is racing along at 105,947 miles an hour. Venus is slower at 78,341 miles an hour. Mars is slower still at 53,979 miles an hour. Jupiter plods along at 29,236 miles an hour and once again Saturn brings up the rear at 21,637 miles an hour.

It was a rare gift presented to us by the cosmos, an alignment that happens once every 10 years or so. The last time it happened was in 2005. That alignment gave new meaning to the term high five and the star studded line up might have been introduced according to size with the biggest first and the smallest last. Jupiter appeared late in the evening followed by Mars just after midnight and so on until the last and smallest planet Mercury appeared just before sunrise which again made it hard to see. The five planets looked like bright stars and the two stars twinkled like distant planets.

If you were in the proper place at the proper time about 45 minutes prior to the show than you saw everything in all of its splendor. The two participants that were back stage were Uranus and Neptune, too far away and faint to be seen by the naked eye. The show in its entirety took up more than half of the entire horizon. The moon also had a cameo appearance, actually a couple of them, first appearing with Jupiter in late January and then with Mars in early February and then with Saturn, Venus and Mercury in the days that followed.

If you missed the show, there was another condensed showing in Mid-August of the same year but Mercury and Venus were hard to see because they were lower in the west during sunset. If you missed all of it, be patient, the next marquee showing of the grand five will be in the year 2020.

Time Traveler!

When we see light from stars that are millions of light years away we don't know if those stars are still alive but we do know that it took that light millions of years to reach us. So in a sense that light is a time traveler that comes to us from the distant past. That same light can also tell scientists the age of the star or planet. Some scientists even argue that star light actually shows signs of age. It's easy to think of star light as just light, or a glow but it is actually a record of chaotic things that happened a very long time ago.

One scientist compared star light to a walker traversing the country side with the distant lights of a city glowing over the horizon. From a distance the hiker only sees is a steady glow above the horizon but the closer he gets to the city the more defined those lights become. Before long the hiker sees head lights from cars, street lights, lights inside homes and businesses and they all combine to make up that one large glowing image that he saw from a distance on the horizon. Researchers say looking closely at star light can reveal the same kind of detail.

Light travels at 186,000 miles per second and can cover 6 trillion miles in a year. If light left a star 10 billion years ago and that light is now just reaching us than that means the universe as well as that star are both at least 10 billion years old. This is why light can be seen as a time traveler.

Here's where things get complicated because light might be able to take short cuts in space. According to one theory, the universe is curved and after studying more than a dozen unique stars, astronomers found that the light from those stars traveled in curved paths instead of straight lines. A straight line is a more direct route than a curved one and the direct route takes less time, so light that comes to us through curved space may tell us more about the universe than light that traveled a straight line.

In the winter of 2003 NASA released a picture of the universe when it was an infant and the headline for that picture read The Oldest Light in the Universe. From that snap shot they determined that the first stars formed 200 million years after the Big Bang, they also determined that the Big Bang happened 13.8 billion years ago.

Astronomers like to use the spectroscopy which is a process that uses light from distant stars to find out what those stars are made of. Light from a star, planet or even galaxy can be passed through a spectroscope which acts like a prism allowing researchers to separate the light into different components and colors.

The spectroscopy determines a star's age by seeing if the star has chemicals other than hydrogen and helium present, if not the star is old because only the first stars were limited to hydrogen and helium. Later generations of stars were able to forge heavier elements so if heavier elements are present then the star is relatively young. All this is gleaned from light. This kind of light does not show us how things are but rather how things were.

For instance, light from the sun takes about 8 minutes for it to reach us and when it does we do not see the sun as it is but rather as it was 8 minutes ago. Light from the nearest star takes about 4.3 million years to reach us and again when it does we are not seeing that star as it is but rather how it was 4.3 million years ago. So astronomers can look into the past when they gaze at stars. When we see our nearest galactic neighbor, Andromeda we see it as it was 2.3 million years ago because that's how long it took its light to reach us. The same is true for the moon, whose light takes one second to reach us.

Using this technique one would think we could see light from the Big Bang, but that's not so. Problem is there was no light at the time of the Big Bang so there's nothing to see. Scientist say during that explosion everything was crushed together with no separation of any kind. Energy, matter, light radiation were all one and photons and light were all part of that super-hot and dark unit. We had to wait a couple of hundred million years after the Big Bang for the universe to cool to the point where there was separation. That's was the moment when light could travel and come to us from across the universe. Any earlier than that and light cannot be used as a tool to measure or calculate time. So if you are wondering how close can we come to seeing the Big Bang, you have your answer.

Old Timers!

Astronomers say they have found the oldest known star in the universe and it is relatively close to us. They say the star was formed about 13 billion years ago shortly after the Big Bang created the universe some 13.8 billion years ago. It doesn't really have a name but they call it HD 140283 and its about 190 light years from our solar system. HD 140283 is made up almost entirely of hydrogen and helium which is a characteristic of the first, old stars. Remember later generations stars are more complex.

How brightly a star burns can determine its age. In this case HD 140283 is starting to dim which means it's probably burning through the hydrogen at its core which can indicate its end is near. Young stars don't exhibit this behavior. There is a star out there that raises the question, how can the egg come before the chicken? The star is called Methuselah and some scientists say it is older than HD 140283 and is even older than the universe. How can that be? In 2013 Methuselah's was estimated at 14.5 billion years old. The universe is only 13.8 billion years old, so again I ask how can that be? Some estimates placed its age at 16 billion years. If that is true how can a star be formed before universe, maybe the same way the egg came before the chicken.

Turns out Methuselah and HD 140283 are one in the same. Who knows exactly how long the two have been going by different names but researchers have been aware of Methuselah for more than a hundred years. It speeds along at some 800 thousand miles an hour and its orbit brings it less than 200 light years from Earth. But the question of how a star could be older than the universe would not go away. Scientist are puzzled but still think Methuselah is 14.5 billion years old, which is still some 700,000 years older than the universe. So they came to the logical conclusion which was there calculations were wrong. So in light of this conundrum scientists say they're committed to further research and more calculations which they say should reduce the star's age down to the point where the results would make sense.

And there is always something bigger, faster, stronger and older and with that we introduce the star called SMSS J03100.36-670839.3. It is a recently discovered star and it might be the oldest yet. Or it might not be. This much we know SMSS J03100.36-670839.3 is in the galaxy. It is about six thousand light years from Earth and depending with whom you speak it is either the oldest known star or the second oldest known star. Its age is estimated to be 13.6 billion years. So are we to believe that SMSS J03100.36-670839.3 has been burning for 13.6 billion years.

And what about Methuselah or HD 140283, thought it that was the oldest at 14.5 billion, but remember scientists who calculated that age admitted that they were probably wrong and need to do more work before they can find the star's correct age, which would have to be less than 13.8 billion years. Only question is when they do come up with the right answer which star will be the oldest?

New and Improved!

The next generation Hubble Space Telescope is called a game changer, no surprise there because you wouldn't expect NASA to call it mediocre. In fact, some people call it the Super Hubble. Expected to debut sometime in the 2030's it should have a mirror that 25 times the size of the one that's on the original Hubble. With the new mirror and other technological advances, the new Hubble is expected to be 100 times stronger than the original and should be able to detect star light that's to too far and too faint to be observed. Another update to the new Hubble is the cost, expected at somewhere between 8 and 9 billion dollars. The old one cost 2.5 billion dollars.

One of the exciting possibilities about the new Hubble is its ability to detect new life on exoplanets. Exoplanets are those that orbit stars outside our solar system. The new Hubble Space Telescope is designed to detect signs of life in new ways like searching for heat. All machines and living matter give off heat. The new Hubble will also look for gasses from waste like oxygen and methane that don't linger long and quickly turn into other compounds. Such heat and radiation have already been detected in a few galaxies that might harbor life.

All this begs the definition what constitutes life! Some of the requirements might be, being self-aware, extract energy from its environment, carry out some sort of metabolism, rid itself of waste, reproduce, create a boundary between itself and its environment. Viruses demonstrate some of these traits but are considered half alive at best. Fire is said to breathe but cannot separate itself from the environment and both are not self-aware. So while astronomers might very well find life it could be so alien that they won't recognize it. At present there are too many planets that are too small for the original Hubble or its successor the James Webb Space Telescope to investigate.

The current Hubble is 25 years old, showing signs of wear and tear and can undergo only so many updates, but still it has surpassed all expectations and has provided invaluable knowledge and insight. If the new Hubble performs half as well it will be an indisputable success.

How long humanity will have to wait to find out if we are alone in the universe is still uncertain, we might never know. What is sure is the new Hubble could provide our best chance for that answer, when considered it is expected to last at least as long as the original which was 25 years and is expected to be significantly stronger.

Astronomers estimate there are at least 100 billion planets in our galaxy and 10 to 20 percent of them orbit stars in what is called the "habitable zone" which means those planets are not too far away from the star nor too close but just the right distance where life might develop. So we increase the chances of finding life by expanding our search and further increase those chances by deploying a superior telescope.

The new Hubble is also expected to identify stars like the sun that are 30 million light years away and in the process deliver the clearest, high resolution picture of the universe we've ever seen. It could even see galaxies billions of light years away with the same clarity that the current Hubble sees which are millions of light years away.

We hope that the future is brighter than the past. Advances in technology have opened doors that were once believed to be permanently closed. As the new generation of astronomers ready themselves for their disciplines they are armed with and benefit from the knowledge of their predecessors. They will almost certainly do better. This new generation will also have superior equipment and technology. Remember how far the telescope has advanced astronomy after it was invented in the early 1600s. It has been argued that telescope has provided humanity with the majority of information about astronomy and without it we would know significantly less. It is hoped that the new generation of space telescopes will produce even greater contributions to astronomy and science than the earlier ones.

What's in a Myth?

The myth of the Man in the Moon has been around for quite a while and bears a bit of clarification. There is no such thing- but the tale might merit clarification regarding how it ever got traction. Historians will tell us that there are several variations of how the story came to be and was passed down through the generations.

Established story in the oral tradition in Europe states that there was a man who was banished to the moon for a crime that he committed. Christians said the man worked on the Sabbath and was stoned to death by God's order. The Romans thought the man had stolen sheep and was placed on the moon as part of his punishment. The Germans thought the man was caught stealing from a hedge that belonged to his neighbor and his punishment was incarceration on the moon. There are many more man on the moon myths but those are some of the more established ones.

In the past many people starred up at the moon and thought they saw some sort of configuration on the surface that resembled the face of a man. Some times that face covered much of the lunar surface and other times it covered just a fraction. But shadows on the moon can be complicated because they are not filtered by an atmosphere. Those shadows come from the Sun, the Earth and the Moon itself which might explain some of those strange shadow configurations.

Here' the abbreviated scientific explanation: there are vast flat spots on the moon called maria and they were formed by lava that filled in old craters and became basalt rock. The near side of the Moon that contains the maria make up the illusion of the man which is always facing Earth. Remember the Moon is tidally locked, does not spin on it axis and always shows Earth the same lunar face. The man's face is on the south pole of the Moon and was detected by a computer that used facial recognition techniques. The computer took a picture of the image which included three large holes situated inside of a circle. Two of the large holes made up the man's eyes and the lower hole made up his mouth, a smaller hole in the middle made up the nose of the man. Down through the ages there have been at least five images that could have been interpreted as a person on the moon and one of those five images could have been interpreted as a woman.

In the summer of 2014 there was a video that surfaced that purported to show an alien like figure walking on the moon. The video went viral. NASA felt compelled to respond and offered an explanation. It stated that what appeared to be a figure walking on the lunar surface was actually nothing more than dust that just so happened to be configured in such a way that it remotely resembled an alien like figure. Another theory stated that the image was the result of a shadow cast by a boulder. NASA's other possible explanation was that the alien like figure was nothing more than a scratch on the negative on the film that recorded the image.

Sightings of men are not restricted to the Moon; some also came from Mars. In 1976 the Viking 1 Spacecraft took pictures of what appeared to be human faces on Mars. The pictures contained the illusion of eyes, mouths and a nose. The image contained what could have been three human faces and at least one partial skull. Speculation was rampant. Some folks even suggested they could have been part of an ancient civilization. Scientists were sure it was just a natural rock formation but by that time it was too late, the media had begun to cover the story and the public's imagination had been captured.

The story almost took on a life of its own, it became a pop icon, a movie and the subject matter for books, magazines, radio and TV shows. The public's appetite had already been primed with the novel

War of the Worlds originally published in 1898 and the 1938 radio drama by the same name that resulted in public hysteria. Given those two events the nation's imagination was stoked with ample interest to make any kind of Martian story a huge national interest. Almost 20 years would pass before there was a legitimate explanation of what the image might be.

It was in1998 when the Global Mars Surveyor took a high resolution picture of the image in the same exact area. That picture was ten times more clear than the first one taken in 1976. It showed that there were no men, no partial skulls and no ancient civilization. Instead it was just a natural land formation created by an optical illusion caused by the Martian surface and its shadows.

History has shown without exception, that given enough time and enough objective observation, science will at some point produce a legitimate answer to all of the perplexing mysteries that have been posed by the natural world. At one point we thought the Earth was flat, another time we thought the Earth was the center of the universe, at one point no enlightened person thought a rock killed the dinosaurs. Science proves and disproves and in this case put the k-bosh on what would have been a really good story.

Gotta Love Technology!

Scientists are giving technology the credit for yet another discovery. They have found the first generation of stars that were thought to be too old and too far away to see and the credit goes to a new and improved telescope that has identified a distant galaxy that contains the old timers which are referred to as population III stars. Population III stars are made up almost entirely of hydrogen and helium which were the first elements present in the universe and continue to be the most abundant.

Again younger stars are made up of heavier and more complex elements which did not form until the early universe had a chance to cool down. According to the prevailing theory, clouds or nebula left over from the Big Bang retained their heat for a while and did not immediately form stars. Heavier elements help the gas cloud cool down and without them, the clouds remained warm and produced stars that only contained hydrogen and helium. Under these conditions those stars grew to enormous size, some 1000 times larger than the sun. Once these big stars turned into furnaces they wasted no time in burning through their fuels. One of these big giants, much larger than our sun might only last a few million years, compared to the sun which can last 9.5 billion years. After the big stars died and went supernova, they sowed the seeds for the future generations of stars to burn heavier elements like carbon, iron and oxygen and whatever else was needed to form life as we know it.

It is theorized that if the universe was 100 years old, the big stars were formed when the universe was six and it's also thought that the first big stars did not form until about 80 to 100 million years after the Big Bang which occurred 13.8 billion years ago.

These big stars were also hot. In the absence of heavier elements like metals, the surface temperatures of the big stars were more than a dozen times hotter than our sun's surface temperature estimated at 10,000 degrees Fahrenheit. We know stars generate heat in the cores but those cores work less efficiently without metal, so as a consequence these big stars had to burn hotter elsewhere in order to counter act the effect of gravity and to prevent themselves from collapsing under their tremendous weight. So they burned hotter and did not last as long as the smaller, second or third generation stars, such as our sun.

A lot of these older stars were found in a distant galaxy called CR7, thought to be the brightest galaxy in the known universe. It could also be one of the oldest because of its Population III stars which contain only hydrogen and helium. Before the new telescope, these galaxies and their stars were only theory but now technology has turned theory into fact.

CR7 is thought to be about 13 billion light years from Earth. Again thought to be the brightest and most distant galaxy in the universe and its stars are thought to be the first, being born just about 800 million years after the Big Bang. It is also thought that galaxies contain maybe a billion or more stars and if half of those stars have the potential to develop life, then what can be hope from CR7?

Well we can't expect to do much because it is so far away. But what it life developed somewhere in that galaxy! Shouldn't that life have the proverbial "head start" regarding its rate of advancement and its ability to communicate and perform long distance space travel? Look how far humans have come in the short time we've had. We've been around for millions of years, 100,000 years as homo sapiens and only about one tenth of that time could we be considered intelligent enough to manipulate the environment. So think of what we accomplished in

10,000 years and compare that to what intelligent life in CR7 might have accomplished in 6 billion years. Would it not be reasonable to expect to hear from these beings, if in fact they had that "head start" and were so inclined to seek out other life in the universe. Question is our we prepared to receive such a visit!

What If!

What would it look like if by some impossible chance a person fell into a black hole? The experts say that would depend on the size of the black hole. Strange as it sounds, the bigger one might take easier on you than the little one but either way the end result would be the same. The event horizon of a black hole is kind of like its door way. When you approached the event horizon of a smaller black hole, which would be about the size of the sun, your body would be stretched. They say the difference in gravitational forces from your head to your toes would rip your body apart. If you were pulled into a massive black hole, its much stronger gravity would take your entire body in at once without any ripping of appendages.

Theorists say the closer you got to the event horizon or door the more it would appear that you were moving in slow motion. It would be a silent motion because remember there is no sound in space because there is no air. All of this sounds like science fiction but the experts say this kind of science, while still theory, is stranger than fiction.

The intense gravity of these black holes, big and small warp the space around them. So as you were pulled inside them you would almost see both sides of the black hole, similar to the image you see in a warped mirror with its center bulging out from its sides. These warped mirrors allow you to see around corners and to a certain extent this is the kind of view you might expect from inside a black hole.

The deeper into the hole you go the more pressure you will feel. The forces of gravity will stretch you vertically and horizontal, long ways and sideways until your body might be thinner than the edge of a razor blade before being torn to pieces. The theories on what happens next are too radical for me to try and relate lest I be accused of drinking and writing.

What are the odds of a black hole entering our solar system and what might happen as a result? The chances of that are really low, estimated by some to be about 1 in a trillion, but if it did happen it would be curtains. If a black hole 30 miles in diameter entered the solar system, we would feel it before we could see it. When it was about a light year out or six trillion miles away, its gravity would start to tear apart the existing order of the solar system. Everything from the sun and planets to moons and asteroids would have their orbits effected. What exactly would happen we just don't know, planets and the sun might collide with each other or some might get ejected out of the solar system, or they might all get drawn to the black hole to an inevitable death. Scientists just don't know; it is a theory. The black hole would not be seen until it had torn apart a planet or two and what was left would start to form an accretion disc around the black hole, the disc would start to glow and that's when you could see the black hole as it reveals its true destructive form.

There is a good chance that the black hole would be much smaller than Earth but its mass would be much greater. It might be 100 times smaller than Earth and a have mass a thousand times greater than our planet. Remember black holes are picture post cards of material condensed beyond our comprehensions. A planet as big as Jupiter could be swallowed by a black hole and while we still could not see the black hole we would know something was wrong by the fact that we just saw Jupiter get torn apart.

Before Earth was torn apart by the black hole's gravity we might experience super earth quakes, monster tsunamis and massive volcanoes. Even though the Sun has a million times the mass of Earth it would not fare much better because it too would be ripped apart by the black hole's gravity. When all was said and done all that would remain is a hot swirling disc of gas and dust. This disc would swirl around the black hole's event horizon and could extend out into space for thousands of even millions of miles. One of the wonders of a black hole is it is a phenomenon that we can't see but are certain it exists because of the gravitational influence it has on material that we can see, making it abundantly clear invisible forces are not restricted to science fiction.

The Men Who Fell to Earth!

If he could have he probably would have jumped from space. Some of the promotions for his high altitude skydive were billed as such, "The Man Who Jumped from Space" was one promotion and while that title was not technically correct it did peak public interest to the point where millions of people watched the entire several hour event live. The main headliner was Felix Baumgartner. He was the young Austrian dare devil who "Fell to Earth" in a record breaking sky dive from what was literally the edge of space. The height at which space technically begins is 62 miles up, he jumped from about 20 miles, still high enough to experience most of the fatal discomforts of space, no air, blistering cold, blood that boils and bodies that bloat, just to name a few.

In the fall of 2012 the high altitude parachute jump was carried live from beginning to end. The jump was also promoted as the man who would break the sound barrier without a jet. The sound barrier is about 770 miles an hour. Many people didn't watch for fear they would witness a dare devil's death wish come true. At an altitude of 126,720 feet or 24 miles above Earth, high up in the stratosphere he stepped out on the porch of his basket that was beneath the balloon, said a few words, something like mother earth I am ready to go home and then jumped. At that altitude there is not much resistance and when he jumped it was liked he was being sucked to Earth. He was gone from view in an instant, not like other sky divers at much lower elevations who gradually dropped out of sight. Felix Baumgartner was immediately gone from sight as if he were sucked down a vacuum tube. He would break several records in one fell swoop, the highest speed, the longest free fall and the highest altitude.

Who could blame Felix Baumgartner if he was afraid, he had his concerns from the beginning. He didn't like his specially made suit because it was too restrictive and didn't allow him the movement he needed to orient himself if he started to tumble during his free fall. His support team had to constantly remind him that the suit was essential because it would keep him alive. Details as to what happened next are not clear but it appeared at one point Felix Baumgartner refused to train with the group that sponsored his jump. So the group trained without him but still used his name on the specially designed balloon and basket. The insinuation was clear; the jump could go on with or without Mr. Baumgartner.

Felix Baumgartner began training for the jump in 2010. Two years later he made the first in a series of practice jumps. His last practice jump had to be aborted because of weather conditions. Problem was he was down to his last balloon. Apparently the high altitude balloons were so high tech that once they were taken out of their boxes they could not be put back inside, so if the weather refused to cooperate again, the jump would have to be delayed further or even canceled. The non-re-usable balloons cost 250,000 dollars a piece.

Here's what his jumped looked like to the observer who watched on TV. It took more than an hour for the balloon to reach its maximum altitude. When the 55 story balloon could float no higher it flattened out like a mushroom as if it were pressed against the ceiling of the stratosphere. Again when Felix Baumgartner jumped, it looked like it was sucked into a vacuum. He quickly reached speeds of several hundred miles an hour. He began to tumble shortly into his fall. He tumbled out of control, head over heels, or heals overhead. He later said he had trouble orienting himself and could not stop his tumbling, He would stick out his left hand to stabilize himself and then the right but could not "catch up." He was falling out of control at some 700 miles an hour on live TV. His top speed was measure at 843 miles an hour and its not known if he could have gone faster had been able to stabilize himself sooner.

For part of the fall it looked as if he might be killed. After what seemed like an eternity, Felix Baumgartner finally stabilized himself and apparently enjoyed the rest of his record breaking sky dive. When he landed he was greeted by cheers and congratulations from mission control, event organizers and his parents who had flown in from Austria. His mother certainly looked worse for the wear, sitting through the several hour ordeal, understandably worried sick that her son might very well die before her eyes as the stress and anxiety were written all over her face.

Felix Baumgartner was an instant celebrity, offers for interviews instantly flooded in from news programs and talk shows whose audiences were eager to know more about this person. And who could blame them, here was the guy who jumped from the edge of space, 24 miles above the ground, fell to Earth at almost 850 miles an hour, and spent 4:20 minutes floating, tumbling and hurtling toward mother Earth, kind of like a real life super hero in an action movie that was popular at the time on the big screen cinemas.

Two years later that record would be broken by an unlikely candidate. The new record holder was not a young dare devil; he was an old man. He was a corporate executive, a computer scientist and he did it without any of the fanfare or media attention of his predecessor. His name was Alan Eustace and when he landed he looked every bit as crazy as the record he just sat. His hair was standing straight up, he looked wired, his eyes sparkled like those of a teenager instead of a 57-year-old man.

So after that jump the sky diving record books were rearranged. Alan Eustice holds the record for the highest jump at 135,890 feet. Felix Baumgartner still held the record for the highest speed at 843 miles an hour and somebody called Colonel Joe Kittinger still holds the record for the longest free fall at 4:36 seconds. You remember Colonel Kittinger from earlier in the book, he sat all three records with one jump more than 50 years ago in 1960.

It appears Felix Baumgartner and Alan Eustice could not be more different in the approaches to their jumps. One was a dare devil and the other is a corporate executive. One made the jump in the prime of his life, the other made it when he was an old man well past his prime. One jump was aggressively promoted in the media and the other seemed to be done in secrecy. One jump was sponsored to the tune of millions of dollars and the other seemed to be funded by one individual. One craft was made to look like a space craft and allowed its occupant to reach the jump height in a pressurized capsule that resembled a lunar craft and the other skydiver had no special capsule but instead rode on the outside on what appeared to be a go-cart with nothing but a pressurized suit to keep him safe. It might be nice if both men could meet and have a chat over a beer and we could privy to that conversation similar to a fly on the wall.

Why Don't the Planets Conform?

Our planets don't seem to adhere to the rules of density that bind our Earthly world. Take for example the reason why ice floats on water, ice is less dense than the surrounding water. Another example would be the reason why warm air rises, because warm air is less dense than the surrounding cooler air. But how can planets float in space when planets are much denser than the surrounding space? If planets were less dense you could see them floating just like ice or warm air but they are denser and should sink to the bottom. But there is lies part of the problem, sink to the bottom of what because I don't think space has a bottom, at least in ways comparable to the way a lake or an atmosphere has a bottom.

One attempt to explain the phenomena states that planets are not floating in space, but only appear to be floating and are really in orbit. If a planet is in orbit around a star that planet is actually getting pulled closer into that star. According to the theory, planets are pulled so fast toward the star that they actually miss the star and end up going around it in a circle like pattern called an orbit. But that does not explain the fact that the planet is denser than the surrounding space which should have the planet sinking but again there is that conundrum, sink where, to the bottom of what? So do planets just sit there, floating?

Another theory states, of course planets don't float but instead are in a free fall and they maintain their orbits because of their forward momentum. But the term "free fall" begs the question where are they free falling to? They are not going to fall to the bottom because I don't think space has one. And yet another theorist suggests planets are not floating in space, they are hurtling through the cosmos at very high speeds but regardless of the speed, planets are still more dense then the surrounding space that they are hurtling through so at some point they should begin to sink. Velocity has also been used as an explanation. Consider if Earth just came to a standstill, where would it go? Our planet is traveling in its orbit around 67 thousand miles an hour and if it stopped moving would it remain where it was or would it drop?

Gravity has been used to try an answer the question but it gives only a partial explanation at best. Two objects that have mass will be pulled toward each other until they collide but they are still floating in space and they are still more dense than the space that surrounds them therefore they should sink.

We consult Newton's first law of motion for clarity. An object at rest stays at rest and an object in motion stays in motion with the same speed and in the same direction unless acted upon by an unbalanced force. If the object in motion or at rest is Earth, it is still more dense than the surrounding space regardless if Earth is moving or it is standing still.

Let's try the Big Bang theory. The explosion thought to have created the universe gave motion to everything in the cosmos and because there is no friction all planets move continuously and they are held in place by the star that they orbit. But this too does not explain why the denser planet does not sink, it explains how planets are held in place in their orbits because everything else is also moving. So are we to believe the reason why planets don't sink in space is because they are moving at high speeds? Well those high speeds don't change the density of the planet or of the space that surrounds it.

Would the planets fall if they stopped moving? No one is really sure if that could happen but if it did there are two leading theories as to what might follow. There would be no more day and night as we know it. One side of the planet would get really hot and the other would get really cold. We also would not have seasons. If the Earth stopped its orbit but continued to spin on its axis conditions would not be much better. According to this theory a day would last a year and the waters on the equator would shift toward the polls because of the lack of a centrifugal force. But there is no mention that a partially motionless Earth might fall, presumably somewhere.

Another theory suggests that if Earth stopped moving the atmosphere would still be in motion at the previous speed of the planet's rotation which is about 1,100 miles an hour. The result would be winds so high that they would scour the Earth's surface to the point where it would be unrecognizable and unable to sustain complex life but still no answer as to whether it would fall. The main reason why there is no answer to this hypothetical question could be all of the bright minds of astrophysics are too busy trying to understand the factual mysteries to be occupied with the hypothetical ones.

They were Second but Still First!

The Apollo 8 mission was the program's second human spaceflight mission but it was the first to leave Earth's low orbit. That means its crew was the first to reach the Moon's orbit and return home. They were also the first to see the dark side of the moon not to mention they were the first to see Earth in all of its glory as a whole planet which is an image called Earthrise. The Earthrise was that iconic picture taken of Earth with the moon's surface in the foreground and about three quarters of Earth is in the background and everything else is covered in the pitch blackness of space. The Earth sat slightly tilted on its axis, brightly illuminated with its blue and white colors against the black back drop. That snap shot has been called "the most influential environmental photograph ever taken." Three men were on board Apollo 8 had plenty to feel victorious about as we will later learn.

The mission had just a few minor problems and was considered the first important step toward a successful lunar landing. NASA did not plan any scientific tests but photography played an important role. The astronauts were to take pictures on the lunar surface for future landings and of course that's when they took the iconic Earthrise picture.

On the way to the moon the crew was also the first to pass through the Van Allen radiation belt which is about 15,000 miles above the Earth's surface. The belt contains significant amounts of radiation and in excessive amounts radiation can be fatal by causing cancer. NASA determined that if the astronauts passed through this belt quick enough it would minimize their exposure to radiation which it calculated as the equivalent of a chest x ray.

Almost 70 hours into the flight the crew went around the dark side of the moon and lost radio contact with mission control on Earth. About an hour later the crew performed a crucial engine burn and if it had been done incorrectly the crew could have crashed into the moon or the men could have been flung into the far reaches of space. Part of that crucial burn was described as the longest four minutes of the crew's lives. The mission lasted about 147 hours, netted the first pictures of Earth from deep space, set a new speed record which at the time was 24,200 miles an hour and provided the first live TV coverage of the lunar surface, not to mention probably the biggest achievement which was the first humans to travel to the Moon's orbit.

We cannot really mention Apollo 8 without mentioning its predecessor or the times of the culture. The year was 1968 and a crew of three astronauts completed the mission. There had been an earlier space flight scheduled for January of 1967 but a fire on the craft killed everyone on board. You'll remember that earlier story about the men killed in that fire and the safety concerns they had before their deaths. They say timing is everything and that rang true when that iconic Apollo 8 Earthrise picture was taken on Christmas Eve in 1968. That was a turbulent year for America in terms or race relations, the Vietnam war and tragic political assassinations.

The astronauts not only showed the Earthrise picture and snap shots of the moon from perspectives never seen before by humanity but they also held a live broadcast for that personal touch. Millions of people watched and listened as the three astronauts set the stage and told their stories as they became the first humans to orbit another world. The astronauts were religious and read passages from the bible and said things like "on the good Earth." It was a touching time, a moment of national pride tempered with national tragedy and uncertainty. It gave hope to many Americans who were losing belief that something special could happen to them and enrich their lives.

1968 was a turbulent year, one that saw a record number of civil rights protests across college campuses and inner cities, it saw the assassinations of Martin Luther King Jr. in April and then Robert F. Kennedy in June, the Summer Olympics saw the first and last ever podium protests over racial equality by medal winning athletes, the Vietnam war was dividing the country, America was in trouble and needed some sort of unified healing. Who knew a small part of that healing would come from three men near the moon on Christmas eve. In fact, those astronauts were told that they would have the largest audience ever to listen to a human voice and it seems that they made the most of that opportunity.

Where Did All That Water Come from?

A new study sheds light on the theory about where Earth's water came from. The study suggests that Earth was bombarded by asteroids and comets both of which can carry significant amounts of water on their surfaces and even more in their interiors. Scientists believe this occurred during the Late Heavy Bombardment Era, a self-explanatory term as to what happened. This period took place some 4.3 billion years ago. But now scientist think that the Late Heavy Bombardment lasted longer than previously thought which makes it more plausible that asteroids and comets had more time to fill Earth's oceans, lakes and rivers with water.

According to the new study the bombardment could have lasted more than a billion years longer than previously thought, and hence the name it was heavy. The Earth is more than 24 thousand miles in circumference with 71 percent of its surface is covered in water, so there must have been a lot of asteroids for a long period to deliver that kind of volume. The Late Heavy Bombardment lasted for billions of years and scoured the Earth's surface. Some of the asteroids that hit were 24 miles in diameter, four times larger than the one that killed the dinosaurs. Earth was hit so often and for so long that the surface didn't have time to cool between impacts and was sterilized when its rock surface became molten.

But could have these asteroids contained enough water to fill Earth's oceans? Consider the asteroid Ceres, the largest one known in the solar system, scientist believe this one single asteroid might contain more fresh water than all of Earth. Ceres is shaped like a sphere and is about 600 miles in diameter. That's about the size of Texas. It's made up of three basic parts, its outer crust, its mantle and the core. The mantle covers the most mass and scientists believe that is where most of its water is stored. Earth has about 41 million cubic kilometers of fresh water. Ceres has about 200 million cubic kilometers of fresh water. Do the math and the argument over whether asteroids brought water to Earth seems to be solved, leaving Earth with an extra 159 million cubic kilometers of fresh water left over from just one asteroid.

Astronomers have since discovered many asteroids that are covered in ice and probably have much more water tucked inside their rocky interiors. They also say this could explain why the water does not evaporate during the searing temperatures when the asteroid enters the atmosphere. When it has crashed into the surface its outer shell of ice had long since melted or evaporated but the water inside its rocky interior was undisturbed as well as other minerals that were essential for the development of early life on Earth.

Scientists like to use the example of an asteroid called 24 Themis which orbits in the Asteroid Belt between Jupiter and Mars. It is covered in a thin layer of ice and contains a lot more water and minerals than what was first suspected. It is about 125 miles in diameter and up to one third of its mass could be water. If 24 Themis is about one sixth the size of Ceres it might be expected to carry about one sixth the water contained in Ceres which would be about 300 million gallons of water, again a huge amount just from one asteroid

But not all water is the same, apparently some is heavier than others. The heavy water contains deuterium which is a hydrogen atom with an extra neutron and is about 10 percent heavier than the rest of the normal water found on Earth. Water from asteroids and comets carry this heavy water which helps to confirm that the water in the oceans came from space. But scientists with the Rosetta space craft beg to differ. You'll remember Rosetta from earlier in the book, the probe that actually landed on a comet. Scientists associated with Rosetta say the water on the comet that it landed on was too heavy to be a suitable match for water found on Earth. Other critics say the water found on the Rosetta comet may not be indicative of the majority of comets that are capable of delivering water to Earth.

There are two types of comets, long term and short term. The long term comets travel far out in space to the Oort Cloud, located trillions of miles from Earth. The short term ones travel to the Kuiper Belt which is billions of miles away. Water from the Oort cloud is too heavy to match Earth's but water from the Kuiper Belt was a perfect match. But apparently all Kuiper Belt comets are not the same and can carry water with different weights, so once again it seemed plausible that asteroids and comets brought water to Earth.

Scientists also believe that shortly after Theia, which was the Mars sized planet that collided with Earth some 4.5 million years ago and created the Moon, asteroids starting pelting the surface and seeding the planet with water and other carbon based minerals that gave rise to life. These asteroids were probably capable of sublimation which is the process where a substance goes directly from a solid to gas and by passes the liquid phase and could help to explain how asteroids were able to retain their water after impacting Earth.

During the Late Heavy Bombardment Earth and other planets were disproportionately hit with large numbers of asteroids, with new research suggesting the bombardment could have lasted until 2.5 billion years ago. Unfortunately, there seems to be no way of finding out how often asteroids impacted Earth during that period. Consider the odds of an asteroid impacting Earth once every 10,000 years, during the Late Heavy Bombardment asteroids probably struck much more often given them ample time to fill Earth with water.

So where did all those asteroids and comets go? The solar system has had 4.6 billion years to sweep itself clean of asteroids. When the solar system was young it had many more asteroids, so much so that it was called a shooting range because of all of the flying debris. But due to gravity and chance, the number of asteroids was reduced to its current number, estimated to be about 150 million that are of significant size.

Asteroids are left over bits and pieces of rock and other debris that did not collect into planets when the solar system first formed. Over time the effects of gravity reduced the numbers of asteroids because smaller objects in space are attracted to larger objects, meaning the bigger asteroids consume the smaller ones. The end result was the bigger objects increased their size but their overall numbers were reduced. Astrophysicists say for every known asteroid there is several more that are not known.

The solar system would probably be home to more asteroids if not for Jupiter, which has been called a giant sweeper. It is estimated that about 3 billion years ago much of what was the early asteroid belt went to form Jupiter. The giant planet's gravity drew close asteroids into its mass. As years passed Jupiter's gravity continued to get rid of asteroids, those that got too close were either flung out into space or into the Sun or again into Jupiter's tremendous mass. So the presence of Jupiter is credited with reducing the number of asteroids to its present count.

The total mass of all of the asteroids in the solar system is less than that of the moon. The largest is Ceres about 600 miles in diameter and the smallest ones are less than one half mile across. There are three types of asteroids C type, S type and M type which are either made up of clay and silicate rock, silicate rock and nickel iron or just iron. So asteroids and comets have mixed blessings. According to scientists they are responsible for our very existence, without them there might not be water and other life supporting materials on Earth and the planet might still be home to dinosaurs. But if one hit now it could undo all the work done by previous asteroids, and cause global extinction, emphasizing how important timing can be.

A Syndrome that Feeds Upon Itself!

There is a syndrome that posits when debris in space collide with each other the impacts create even more debris and with the passage of enough time the amount of debris in Earth's low orbit will be too cluttered for normal space operations to continue as they have in the past. Earth's low orbit extends about 1,243 miles above the surface. The amount of debris left in this orbit could have serious consequences for objects like satellites or the International Space Station which include Earth based communication, smart phones, entertainment and security functions, The syndrome takes its name from the man who came up with the concept, NASA Scientist Donald Kessler in 1978. The Kessler Syndrome uses terms like "domino effect" and "feedback runaway" to describe the syndrome. Scientists before Mr. Kessler kept records of objects in space but did not update those records after those objects collided with each other. Mr. Kessler took it one step further and not only pointed out the present danger from the existing space debris but also pointed out possible future collisions based on orbital trajectories. He then suggested a remedy: reduce the amount of space craft in orbit that was not functional.

These objects in orbit are flying by at amazing speeds around 17,000 miles an hour per hour and there are thousands of them big and small, the largest thought to be 2,000 pounds. All of them are subject to drift and decay and none travel a straight path. Variation in Earth's gravity causes drift and the friction from Earth's atmosphere causes decay. Both make it more difficult to track exactly where this debris is going and how fast. This has direct consequences for dead satellites, which are ones that are no longer operational and just drift in orbit.

Some scientists say these dead satellites pose the biggest threats and compare them to "sitting ducks" because they cannot be maneuvered to get out of the way of another object and they all threaten to create more debris after they are hit.

Compounding the problem is that fact that Earth's orbit is getting more cluttered. There's a lot of junk up there, maybe more than a million pieces in all, 500 thousand of which are larger than four inches. Imagine the kind of damage a four-inch bolt could cause traveling at 17,000 miles an hour. Monitors on Earth are currently tracking some 300,000 pieces of orbital debris. As we've already learned collisions happen and threaten to occur again. Here's one scenario: colliding debris creates more debris and more debris creates more collisions and more collisions creates more debris, you get the picture.

There have been a number of movies based on the problem and while none are entirely accurate they are entertaining and more importantly they bring the problem to the public's attention. One of the more popular stories involves American astronauts in orbit and while making repairs to their space craft they encounter problems. First there is an explosion on a Russian craft and that debris stays in orbit and eventually circles around and crashes into the Americans, killing all but two of them. The general premise of the movie seems to be correct but some of the details were not, one that comes to mind: the orbiting space junk was not traveling near 50,000 miles an hour but more like 17,000. Still that was more than fast enough to create the same kind of damage and injury that was shown in the fictional movie.

European scientists have chimed in on the problem and propose removing a minimum of five to 10 of these large objects every year. That could be unlikely considering the costs that such a removal project would place on space programs that already have restricted budgets. And that's not to mention these space programs don't really have the cost effective technology for successful removals in the first place.

Remember these programs were primarily concerned with putting people and machines into space with the intent of only returning people back to Earth. What to do with the machinery still left in orbit was not even an afterthought until recently. There is certainly nothing like an orbital dump truck that can clean up the place by making the rounds. Things like lasers, tethers and devices that enhance drag have been considered but their applicable technology might be so underdeveloped that it is not even an option. So apparently as on Earth it will also be in space, humans have a hard time cleaning up after ourselves where ever we go.

Houston we have a Problem!

The Hubble Space Telescope was launched into orbit in 1990 and still remains on the job more than 25 years later some 341 miles above Earth. The telescope is 2.5-billion-dollar joint venture that involved the Americans and the European Space Agency. It is an observatory about the size of a bus with a mirror that is 8 feet wide and can see infrared and ultraviolet types of lights. It is also credited with changing the way we look at and understand the universe.

It was named after the accomplished American Astronomer Edwin Hubble who came up with something called the Hubble Law. Part of the law basically states that a galaxy's velocity of recession is directly proportional to its distance from us or put another way the speed at which the galaxy moves away from us can be compared to how far it is from Earth.

But like most things the Hubble was not without its problems. The first plans for the telescope were delayed for years, maybe as far back as the 1940s. The idea was revolutionary and the cost for a space telescope was out of this world. It wasn't until the 1970s that congress finally appropriated funding. Then after 15 years of relative trouble free preparations the Hubble launch was delayed by the Challenger disaster. That explosion happened in 1986 and killed all of the astronauts on board. NASA grounded all of its flights for almost three years until it could figure out why the Challenger exploded. The Hubble depended on those shuttle flights for its launch and regular maintenance. Hubble was not ready for launch until 1990, it was seven years behind scheduled and more than a billion dollars over budget.

To err is human and humans made the Hubble which is an appropriate opening line for the next bit of information. A short time after that 1990 Hubble launch scientists were able to view the first pictures from their crown jewel and they were surprised, in a bad way. They were devastated to find out that the mirror did not work correctly and the pictures were out of focus. What good is a picture if you cannot make out the image? In all fairness those first pictures were no better than the ones taken from Earth which completely defeated the purpose of the entire program. Turns out the main focusing mirror had been polished to the wrong specifications. This was not the news NASA wanted to hear, remember it had already made one big mistake when it erased the tapes from the first moon landing, not to mention the Shuttle disaster and now this, it made another mistake with its big, new, expensive space telescope. The miscalculation on the polished lens was very small, less than $1/50^{th}$ the width of a human hair. Some people called it the small mistake with big consequences.

The media picked up on it and had a field day at NASA's expense, some even calling it the 1.5-billion-dollar blunder. One NASA director had to explain why the program has invested 10 and a half years and spent 3 billion dollars on a project that didn't work properly. Apparently some of the top engineers were sitting at their desks, getting drunk in the middle of the day, racked with stress and disappointment. And when the chips are down the blame is passed, finger pointing went from a technical failure to a leadership failure and even a failure of the culture that was associated with the project. NASA would have to wait three long years for a remedy, when a crew of space walking problem solvers installed a set of miniature mirrors to correct the problem. After that NASA did not have to wait long for its crown jewel of a telescope to start living up to its potential.

The list of the Hubble accomplishments is too long to list here but here are a few: in 1995 it was responsible for producing what is called the 'Hubble Deep Field" which is an archive of never before seen pictures of the deep universe. It contains images of more than

5,000 galaxies and some are as far as 13.2 billion light years away. Remember when telescopes look at stars they are actually looking back into time and in this case all the way back to the near beginning of the universe, believed to be 13.8 billion years old.

Hubble only takes pictures in black and white so astronomers take many exposures of the same object and then go back and lay in different colors such as blue and red in order to capture different wave lengths which brings out the detail in pictures that otherwise would go unnoticed.

The Hubble is so powerful that it could see light from a fire fly 7,000 miles away. That's almost the distance from New York City to Hong Kong, looking across the United States, past all of the Pacific Ocean to the main land of Tokyo to see a fly that could fit comfortably on a person's finger nail. Its viewing accuracy is unparalleled and has been compared to shining a laser beam on a dime several hundred miles away. While it's about the size of a school bus it was fitted into the cargo bay of one of the shuttles with almost no room to spare, which might have caused a moment of anxiety for engineers if they had to explain how they overlooked something as fundamental as the telescope was too big for the trip.

The Hubble tells us something new about everything it sees, including stars that are just forming to stars at the ends of their lives to all of the developmental stages in between, all the while producing new scientific information that was previously unavailable. One such example was when it discovered an extra solar planet. That's a planet outside of our solar system that orbits its own star. It also found many other planets by identifying the changes in their atmospheres as they pass in front of their stars which was then recorded by Hubble. It has also created a map of dark matter, which is a huge achievement considering dark matter is thought to make up the much of the universe.

It discovered two moons of Pluto which are about 46 billion miles from Earth. It helped to determine the speed at which the universe is expanding and also discovered almost all of the major galaxies that are anchored down by a black holes and at the same time helped to locked down the age of the universe. It also found the first evidence of super massive black holes and in the process settled many old debates that divided astrophysics in the scientific world. And anyone can make a request to use the Hubble which of course does not mean anyone can use it, just that anyone can ask.

Since 1990 it has made more than a million observations of space, both near and far. It has provided enough information, data and pictures for scientists and other writers to publish more than 12,000 papers. It has been called one of the most productive and prolific scientific instruments ever built. And while it does not actually travel to distant stars and planets, it takes the highest resolution images of these places humanity has ever seen. It does this while speeding along in its orbit at some 17,000 miles an hour again 340 miles up. While on the job Hubble has traveled about 3 billion miles. It has no thrusters, so in order to change its view point it spins in wheels in the opposite direction to achieve a new viewing point.

Despite it significant size the Hubble uses very little energy. It typically uses about 2800 watts of electricity. The common appliances in a kitchen might only use 2200 watts. It has been called a giant digital camera and despite all of its accomplishments, it cannot look into the sun because the bright light would damage its lens. It also cannot look at Mercury because Mercury is too close to the Sun.

An unexpected benefit of the Hubble was the silicon chips used to look into space were also adapted to detect early breast cancer in women by identifying tumors that could be cancerous and distinguishing them from those that are benign.

Could the many accomplishments of the Hubble Space Telescope make up for the disappointments of NASA's human space flight program? The Hubble has certainly delivered much more than what was expected and the opposite might be said of our human space flight program. 30 years ago we would have expected to have accomplished much more in terms of space travel, we would have expected to maybe establish a colony on the Moon and have already made the journey to Mars. We're not even close to either. 30 years ago we had no idea we could confirm that the universe was expanding and contained super massive black holes and that there was all of this dark matter and dark energy. So while the Apollo and Shuttle programs exposed our inabilities to travel in deep space the Hubble Space telescope revealed what we could look forward to if we ever learn how.

So how will the Hubble die? Apparently there is one of three options. The first is to take it out of orbit and it will eventually come back down, possibly somewhere over the Pacific Ocean. The second would be to boost its orbit to a higher elevation which would keep it in space for decades and the third would be to do nothing and allow the Hubble to come crashing back down to Earth in an uncontrolled fashion. NASA states the third option is not an option. The second option sounds like the most viable, which would involve sending the Hubble to a higher orbit for a decade which would allow us more time to solve the problem or as one scientist put it, allow us to put off the problem until someone else has to deal with it.

What Could Have Been!

The Constellation space program was an ambitious project that hardly got off the ground and certainly never reached it's potential. It was NASA's latest program that focused on human space flight. It was supposed to replace the Shuttle program but Constellation didn't live long, from 2005 to 2009. There was considerable disappointment over the cancellation of the program which never achieved its goals, some of which were, to return to the moon by the year 2020 and to reach another planet.

Constellation was created to fulfill the dreams of NASA administrators and one former President. The majority of those dreams could be laid out in three steps. The completion of the International Space Station was the first, returning to the moon was the second and going to Mars was to be the third, with the three stages illustrated on the program's logo.

The second goal of returning to the Moon was a little more complicated, apparently the Constellation program did not suffer from a lack of ambition. Program directors were not satisfied with sending astronauts to the lunar surface for hours or even days but much longer, possibly weeks or even months. Plus, the astronauts would need lots of supplies and space ships would have to be built to carry all of the essentials. New tools and technologies would be required. Crews would have to build semi-permanent shelters where they could live and work. They would need facilities were they could store food for long periods of time and would have to have some sort of structure that could deal with minor medical emergencies.

That's a lot of stuff. Of course all that material could not be carried with the astronauts. The typical shuttle could only carry 7 astronauts and a payload of about 50,000 pounds which is the equivalent of about 13 average sized cars, that's a lot of room but still not large enough. NASA would need two vehicles for the job, one for the crew and one for the supplies so it developed two rocket boosters, Ares 1 and Ares V. The crew would ride in a capsule called Orion which would sit on top of Ares 1. Ares 1 was to be the largest booster in the world, taller than a football field is long. Ares V was the rocket that would have launched the crew's supplies. It would have been even taller and could have carried 207 tons to a height of about 100 miles which is in earth's low orbit and from there it would have hauled almost 80 tons all the way to the moon. That's the weight of two average American homes, across nearly 240,000 miles of space.

NASA was spending 35 billion dollars to build Ares I and Orion which could carry up to four astronauts and it was all supposed to be done economically, safely and within a budget. The program was scheduled to launch in 2023. The events for Constellation were to be scheduled as follows: in 2017 or 2018 it would have traveled around the moon in an unmanned test trip in preparation for a voyage to a planet. In 2020 a trip to an asteroid was planned. In 2030 a trip to Mars was supposed to be scheduled which included astronauts on board. It's not likely that those plans will come to fruition, there's not enough money or motivation for the first two and we really don't have the technology to complete the third.

Here's some background about Constellation's beginning and ending. In the winter of 2004 President Bush came up with an idea for humans to return to the moon and he wanted to get there by 2020. So the President gave a speech which basically directed NASA to come up with a program to meet the objective. The speech was called the Vision for Space Exploration which grew into what was called Project Constellation.

Five years later when President Obama took office in 2009 he ordered a review of the human space flight program and the results would not be good. Apparently it was not in the budget, literally, stating that Constellation was "not viable" under the FY 2010 budget plan. The next year the President canceled the Constellation program and during one of those rare moments, congress agreed.

So here's where we stand, 24 men have gone to the moon, half of them have walked on the lunar surface and three have even been to the moon twice. Since 1958 the United States has spent significant tax payer funds on space exploration, much of which was focused on getting to the moon and possibly even farther to a nearby planet. That's more than a half a century and billions of dollars spent on a single objective that is suddenly no longer "in the budget."

So Long Ol' Friend!

So what do we do now without the shuttle program? After 30 years can the US space program no longer transport its astronauts to and from space? Well according to NASA, the agency has not left the business of human spaceflight. The administration is considering plans to construct a large rocket capable of heavy lifting that can transport crew and cargo to outer space destinations but it is not expected to make a decision on those designs for quite some time.

In the mean time we can rent. That's right apparently NASA has worked out an arrangement with the Russians to rent space on one of its vehicles called the Soyuz Spacecraft. The trips would take astronauts to the International Space Station. The price per ticket: 63 to 81 million dollars a seat. That's right, up to 81 million dollars! Where are the Americans going to find that kind of money in our budget when there wasn't enough money to continue with our previous space program that we had to drop?

Anyway, NASA says the arrangement is only temporary but won't state how long the current situation will last. If it's any consolation this is not the first time NASA has been without such capabilities. You will remember there was a six-year gap from the last Apollo flight to the first Shuttle launch. During this latest gap NASA will try an encourage private aerospace companies to step up and help by offering commercial space flights, some presumably to the public in a few years. The administration also has other plans. It will rely on robots from Russia, Japan and Europe to maintain some of our American responsibilities on the International Space Station. The bottom line: for the foreseeable future NASA won't be able to send its astronauts into space.

For the last 30 years the shuttle program was really the driving force behind NASA and now what? Rely on other countries' robots, rented seats and help from a private industry that so far has not helped much. What are the repercussions for NASA? Some critics say what will follow is a huge "brain drain" where NASA's top minds will leave the administration after the discouraging developments of the recent past. That discouragement could quickly spread to the public and with no more human space dramas to get excited about, the American sentiment toward the space program could dry up even more than it has already. Remember space exploration is only important when it's relevancy is applied to humans.

Consider the ugly facts about the Shuttle program that some folks might bring up in order to justify its elimination. More people have been killed on Shuttles than on any other space vehicle in history. Seven people were killed on the Challenger and another seven were killed on the Columbia. Put another way out of the seven Shuttles two were destroyed by fatal flaws. That's a 40 percent failure rate. If a commercial carrier had a failure rate one tenth as high, it might be grounded permanently. Every time a shuttle successfully landed it beat the odds presented by the inferno of reentry, illustrated by the fact the shuttle would sit on the landing strip, too hot to touch, needing considerable time to cool down from the 2300 degree temperatures it endured, successfully running the gauntlet all but twice. But with the passage of time now providing the advantage of clarity, it seems in order for the Shuttle to be truly successful it had to run that gauntlet without a single error.

Was the shuttle too expensive for what it delivered? An estimated 355 people rode the shuttles during the 30 years of the program's existence and that total cost was estimated at more than 170 billion dollars. Is that a good return on the public's investment? Given the vast expanse of space there are people who say the shuttle never went very far. The Shuttle was only designed for low orbit travel, anywhere from 99 miles to 1,243 miles above the Earth. That's just barely scratching the surface when one considers the distance from Earth to the Moon is 238,000 miles and that too is still just scratching the surface.

It was also thought that each shuttle would fly up to 50 missions per year but in reality they only flew about 4 flights per year. So did they meet expectations? And there always seemed to be the thought that the shuttle was going to be replaced by something better, but that was before President Obama took a look at the Constellation program, once thought to be the shuttle's predecessor and canceled it with congressional approval.

Some critics say due to its design the shuttle could only achieve so much and was sort of trapped in low orbit. It completed more than 130 missions over a 30-year period. From April of 1981 to July of 2011 was the official time frame of its life span. In 1983 it carried the first woman into space and later that year carried the first Black man into space. In 1984 it helped the first object to be retrieved and repaired in space. In 1990 it deployed the first space telescope with the launch of the Hubble. In December of 1996 it completed the longest shuttle flight of 423 hours. In December of 1999 it made repairs on the Hubble and in October of 2000 it flew its 100th mission.

The immediate future of human space travel for the United States is certainly uncertain and in no way looks as bright as it once did. Problem is there is not enough public interest and where there is no interest there's usually no funding. When people are worried about finding jobs, paying bills and even putting food on the table, it's hard for them to get motivated over expensive space programs that have no direct relevance in their lives. If you could change the latter problem the former might solve itself.

Why Do We Need One?

With so many costs associated with space programs and all of the inevitable fatal mistakes that come with those developments many people wonder why do we even have them? Do we need them? One of the most common critiques of space programs include something like, we need to correct the many problems on Earth before we go gallivanting around the solar system. Others say tax payer's money should be spent on tax payer's problems and space exploration is not one of them. There are probably a dozen arguments that use some form of this reasoning and they are all hard to argue with because they are all sound.

But if one steps away from the pragmatic approach to the question and takes a philosophical approach the answer might be different. There are those who say we must have a space program because it is in our DNA. They say one of the most basic of human traits is our innate need to explore. They also say the need to explore is what makes us human. We are naturally curious and ever since we became homo sapiens we have been almost obsessed with learning as much as we can about our environment and where we fit in. We've been wondering where we came from since we first started wondering and many people believe that those questions can only be answered through space exploration.

History is punctuated with such examples. What about the first humans on the African savannah who traveled north into Europe to explore a new world! What about the Europeans who traveled over sea to explore the new world, including our nation! What about the explorers who came across the North American plains to the west coast! What would happen if those explorers never embarked on those adventures, chances are our lives would be vastly different? Just think if we never had a space program, again chances are our lives would be different. Explorations from the past, including those in space have delivered huge scientific and economic benefits that were not even imaginable when those voyages were undertaken.

Some people even compared an existence without a space program to spending one's entire life in the same small town, oblivious to all of great wonders, big and small, trials and tribulations, successes and failures that make up the world outside. They might also say we need to explore space so that we can properly develop as a society which is as essential as regular exercise and proper foods are for developing children.

Then comes the argument for space exploration based on resources. A civilized society needs resources. An advance society that is growing and expanding needs even more resources. At some point we will exhaust Earth's resources and you probably know what will happen next. When resources grow scarce there's competition which can and does turn violent. We are already a violent people who have history of killing each other for material gain. You get the picture. Water is one of the valued of resources and it has been discovered in tremendous amounts in space. The natural resources that are limited on Earth might be forever replenished from things like asteroid mining. Sounds like science fiction but could become established science by the year 2020. Water from asteroids could be transformed into fuel for rockets as well as sustain life for humans. The minerals from asteroids, such has cobalt, iron and nickel all have valuable and essential uses.

What about something called survival, that little old, most basic of a human needs, self-preservation. While it's true we don't need to get off the planet anytime soon, we will at some point have to go.

Nothing lives forever and the same is true for Mother Earth, with a life expectancy of about 8 billion years. That leaves us with another 4.5 billion years to figure out how to get off the planet and colonize other places.

Not only does Earth have a shelf life, there is also the threat of asteroids which can cut that shelf life dramatically short. At one time it was considered a ridiculous theory, now it is widely accepted that the dinosaurs were wiped out by an asteroid no more than six miles wide. The Earth is 24,000 miles in circumference, a six-mile-wide asteroid is like a microscopic pin prick that caused global devastation. It took out the dinosaurs in cosmic instant after those big beasts had dominated the planet for more than 100 million years. We humans threatened the planet with irreversible harm after scuttling around on the surface for just 200,000 years and the last 6,000 have been when we are so called civilized. So humanity would be naive to think we are long term tenants on the planet.

Even countries such as Indonesia, Pakistan and Malaysia whose fiscal budgets are much less than America's have space programs, which might speak to the importance of space programs to tax paying citizens as well as those who can't afford to pay taxes. Then there are third world countries such as Ethiopia, a struggling, improvised nation dealing with internal strife, wide spread hunger and disease but yet is has the beginnings of space program. Ethiopia's in dire need of the basic requirements but it recognizes the need to spend about 3 million dollars on a space observatory, which is the first step toward a legitimate national space agency. Could a national space program be a basic need for healthy mental conditions for its citizens? It has been said that poor countries want space programs more than rich countries, possibly because of the mental aspect and again the hope and the ability to dream about space and the sense of national pride that comes from this.

Britain does not have a human space program nor does it contribute any resources to the European Space Agency. In fact, the United

Kingdom, including Wales, Scotland, and Northern Ireland all take the same position of non-participation in human space programs, citing costs and the inability of collaboration. They have instead focused their efforts on satellite technology.

While satellites don't make the world go around they certainly help with the processes. Here's how many satellites are in space and some of their benefits. At last check there are about 2,271 satellites in orbit with Russia launching 1,324 followed by the United States at 658 and the remaining 1,613 satellites belonging to other nations or private companies.

Satellites enhance communication from telephone, television, navigation, business and finance, to monitoring the weather, the climate and environmental change to national defense to monitoring space and the changes in Earth. 50 years ago we did not have satellites but since then we have become very reliant on them and they have made our lives more comfortable, interesting and rewarding.

Some satellites are a few hundred miles above Earth, others are 25 thousand miles up. Data from satellites combined with computer models provide valuable insight about Earth and space that otherwise could not be gathered. Below are a few examples: One satellite tells scientists why Earth's polls are so cold in part by measuring the Sun's impact on our oceans. It is about the size of small school bus and circles the planet 14 times a day at an altitude of 500 miles up. 22 thousand miles up is a chain of five satellites called GOES East. It is a geostationary satellite comprised of five satellites that always point at the same five spots on Earth while the planet orbits providing unparalleled information about the planet and its weather. It gives scientists a complete view of the entire Earth, 24 hours a day on how energy moves around the globe beginning with sun light interacting with water.

Another satellite can look through clouds with a microwave scanner to monitor changes in the ocean and can take measurements of sea ice in places where humans cannot go. And yet another one can detect the shape of the ocean floor miles below the surface. This satellite is about 500 miles up and is accurate within a half inch when viewing earth, enabling it to clearly see the color of a person's eyes 500 miles away.

NASA's TRMM satellite short for tropical rainfall measuring mission satellite carries a high speed camera so fast that it can record individual lightning strikes as well as the distribution of lightning over the entire globe, recording 40 strikes every second which add up to more than 3 million lightning strikes a day. The Solar Dynamics Observatory is another satellite orbiting about 22 thousand miles above Earth, pointed away from Earth and toward the Sun to monitor what is called space weather, known as radiation from the Sun. It does this around the clock by looking at light that is beyond the known spectrum.

Those are just a few of the many satellites and their uses and we've not gone into detail about the human dependency on computers, smart phones, global positioning devices, and the different forms of social media that dominate so many lives; not to mention the wide variety of security functions on a personal, commercial and on a national security level. One can only imagine the silence, isolation and considerable inconvenience that would burden humans if just a portion of those satellites stopped working.

The current fleet of satellites is aging and by some estimates NASA could lose about half of its satellites over the next ten years which could significantly reduce the incoming data that makes NASA so valuable and relevant in enhancing our lives. Compounding the situation is the problem of space junk, up to one million big and small pieces of old rockets and satellites left behind in space threaten to make Earth's orbit too cluttered to launch large numbers of new satellites. So while satellites certainly have problems, their benefits are clearly established and in light of being incapable of significant space travel we might want to place new importance on this technology.

Chill!

The universe has been doing a lot of things for a long time, including cooling since the very beginning, estimated to be 13.8 billion years ago. It has also been expanding for a long as it has been cooling, safe to say the more it expands the more it cools. We can measure the temperature of the cooling universe by something called the Cosmic Radio Background or CRB.

The CRB is the glow left over from the Big Bang which has the tendency to keep gas warm. It is this radiation left over from the Big Bang that fills the universe in every direction. Its microwaves are invisible to the naked eye and can only been seen with high tech instruments. CRB represents the earliest radiation that can be detected and has been compared to light from the sun that penetrates clouds.

There are three ways to transfer heat, conduction, convection and radiation. Heat up one side of an object and the other side also gets warm, called conduction. Air that is circulated is called convection and when heat is transferred in the vacuum of space it is called radiation. Think of how an old radiator works, the metal heating system radiates heat out into the room on a constant and even level. Radiation is the primary method of transferring heat in space.

Apparently when it comes to extreme heat and cold, we can achieve on Earth what the universe cannot. According to scientists the coldest temperature in the universe is 455 degrees below zero. This temperature has been called absolute zero, where temperatures can go no lower. It is the point where there is no heat energy left in anything like atoms to be used for work or function. Atoms are attracted to heat, the higher the temperature the faster they move, the colder the temperature and the slower they move. Apparently the atoms are motionless at absolute zero because the extreme cold has robbed them of their ability to move. But in a laboratory scientist were able to conduct experiments where they were able to reduce temperatures down to 459 degrees below zero. That is 4 degrees below what was previously thought impossible. You'll remember other scientists working on the Large Hadron Collider were able to reached temperatures of 7.2 trillion degrees, much hotter than the core of the sun and hotter even than those left behind from the Big Bang, or any other known temperature in the universe.

So it seems a bit ironic that we consider one possible end to the universe that results from a really cold condition possibly caused by the aforementioned expansion referred to as heat death. Heat death is a theory or concept of the universe that deals with entropy and closed systems. Entropy is defined as the thermodynamic quantity that represents the unavailability of a system's heat energy that can be converted into mechanical work.

Some scientists theorize that when entropy ceases to exist so will the universe. In the most basic of terms entropy is when a system cools down to the point where it no longer has any heat driven energy to do anything. This happens when all of the heat in the universe moves from hot and warm places to cool and cold places. When there is no more heat there is no more energy generated.

We are told that the same law applies to all forms of energy. According to theory, the universe can get no colder than absolute zero which when measured on the Kelvin scale is different than the common zero used on the Fahrenheit scale. Again the maximum low thought for the universe was 455 degrees below zero but conditions created in a lab were even lower. The Kelvin scale was developed by British inventor William Kelvin in 1851. Part of the reasoning behind its creation was the temperature range in space was too radical to be most efficiently gauged by the current methods of Celsius and Fahrenheit. The Kelvin scale also does not use negative numbers and seems to be preferable when it comes to measuring extremely low temperatures. The lack of negative numbers also makes it easier to determine the differences between extreme temperatures. Scientists say everything that happens increases entropy or in other words every movement no matter how small puts off some amount of energy. The universe is too big for humans to comprehend and there are so many moving parts in its great expanse. So as long as there is some sort of moment in the universe, heat will be produced and the universe will be able to stave off this notion of heat death. Scientists don't know when the universe could die from heat death, maybe a trillion years or longer.

How Would You Like to Go?

What better time to talk about ''ways to go'' then following the theory of heat death! Many people are familiar with the Big Bang theory on how the universe will end. Have you heard of the Big Crunch theory? It's basically the reverse of the Big Bang where everything in the Big Bang expands outward from a small inner starting point.

All of the expanding matter than makes up the universe at some point will stop expanding and in theory begin to contract. The driving force behind the contraction would be gravity. At first the contraction would be subtle, the universe will continue to expand outward and then gradually slow down. Under the force of gravity, the expansion of the universe will continue to slow until it stops and then the gradual process of contraction would begin, soon to be accelerated almost exponentially to the point where everything is contracted right down to one very small and very dense point. At that moment everything in the universe is gone, all of it just got compressed into a very small area thought to be no bigger than the head on a pin. That's the Big Crunch theory and through the process of expansion and contraction the universe might live forever

Then there's a different kind of a black hole theory. Instead of black holes eating a planet, this one would consume the entire universe. Here's how it works. Most matter in the universe is thought to be orbiting a black hole. Galaxies like the Milky Way are thought to have these black holes at their centers. Given enough time, probably a couple of billion years these black holes eventually consume their galaxies. After enough galaxies are consumed what would be left would be a disproportionate number of black holes. These black holes would greatly increase their sizes because the larger ones consume the smaller ones. This process would continue until maybe just one black hole is left.

Apparently black holes lose their mass over time, imagine something with gravity so strong that not even light can escape, undergoing such dramatic changes that it can no longer hold on to its own mass. As this massive black hole kind of evaporates it would leave behind subatomic particles that would be distributed evenly. But to the common person nothing would be left behind because subatomic particles are far too small for us to recognize or even really comprehend.

Then there is the Big Bounce, which is similar to the Big Bang and the Big Crunch but the ending of the former is a little more palatable. We begin with the same basic state as the other theories in that the universe is expanding outward. Then gravity starts to slow the expansion and eventually it comes to a halt and then reverses itself into a contraction. Over time the contraction process is accelerated to the point where there is a super rapid and super strong compression which is significant enough to start up another Big Bang. Under this model the universe is not so much destroyed as it would be recycled. That's the Big Bounce theory and it also seems to suggest that the universe might live forever.

Some scientists don't like the Big Bounce theory because there is no way to determine if the contracting universe was compressed all the way back down to the point of the Big Bang. What happened if the universe fell just a few degrees short of the total contraction required to ignite the Big Bang? Would there still be a Big Bang? Maybe there would be just a small bang or even something called a partial bang? Either way if the universe did not contract all the way back to the single unified point, according to the theory it would be repelled by a cosmic force called the Big Bounce. Like the Big Bang, the Big Bounce would produce a new universe but there is no way of telling if that was the second time a new universe was created or the 222nd time.

Here's where things get even more strange. There's the theory called the Vacuum Metastability Event. This theory suggests that the universe is not stable and at some point might blow up or break apart or dissolve or just plain do something bad. According the theory, a bubble appears while the universe is trying to become stable. The bubble could be compared to an alternate universe. It is initially small but starts to grow and then expands in every direction until it has consumed the entire universe. So after the original universe has been consumed there would be an alternate universe. Now there are two universes, at least according to the theory. The law of physics would probably be changed as well as other laws. Scientists are not sure if life would be able to survive or if new forms of life would emerge.

The End of Time theory is stranger yet. This theory suggests that one-day time will just stop. That's right the eternal time, that's been around for at least the Big Bang, 13.8 billion years ago maybe even longer, just stopped. There would be no way of telling one moment from the last, there wouldn't be any more moments, time is frozen, forever. You'd never know that time had stopped or that anything was wrong, you would be frozen in an instant that would last forever. You would never die and never grow old and never be aware of anything. That sounds a lot like science fiction, but so do so many other theories.

Then there are a couple of theories that suggest the universe will not end. One is the Eternal universe theory and the other is the Multiverse theory. According to the Eternal theory the universe is just that, eternal, everlasting with no real beginning and no ending. That almost sounds like something out of the Bible, but it's not and it goes like this. Unlike the Big Bang which is credited with starting time, the Eternal theory suggests that universe is in a cyclical pattern where it has been expanding and contracting for ever, long before the existence of time. The theory also posits that time existed before the Big Bang and could have been created out of a collision with something called branes. Branes are thought to be sheet like structures in space that are formed and exist on a different level than other life as we know it. It is thought that these branes leave traces of their residue in the background radiation of deep space. While it is just a theory, it could be something more in 20 years which is when results from a satellite that is studying the matter is expected to produce data on the theory.

The Multiverse theory is just like the term suggests, there are multiple universes out there, thousands, millions or maybe even billions of them. Or stranger yet there is an infinite number of universes and each one is a little different from the rest. Under a theory like this it is easy to see how a universe with an infinite number could live forever. Problem with this theory is something called physics. Apparently the laws of physics don't work with a multiverse that is infinite in nature. Physics requires something real or finite to work with, some sort of boundary, a tangible line beyond which nothing can expand. Physics does not always agree with many theories which leads some scientists to say physics is where the rubber meets the road in astronomy.

No Easy Walk!

No one ever said a spacewalk was going to be easy. In 1965 Russian Cosmonaut Alexey Leonov found that out the hard way. He encountered problems on the first space walk in history and if not for his fast thinking it might have been his last. Mr. Leonov was outside the craft conducting experiments when he discovered a problem. It cropped up when he tried to reenter the space craft. Seems his space suit became over inflated in the zero pressure environment of outer space. The space suit was so inflated that he couldn't get back inside the ship and was soon to be at risk of dying in space. Since this was the first spacewalk, scientists and astronauts really didn't know what to expect. Turns out there was a pressure relief valve on the suit, which he used. The suit became deflated. The cosmonaut was able to get back inside without further ado. End of story, no, actually it is the beginning of another almost identical one.

This tale is called ''The Spacewalk from Hell.'' The year was 1966 and the Gemini 9 astronaut was on his third spacewalk when he encountered trouble that could have made it his last. He was outside the space craft conducting experiments when he started to get hot. His helmet started to overheat and he felt exhausted so he tried get back inside but he could not fit through the door. Sound familiar! When he pulled himself up to the hatch he found that his space suit had inflated to the point where he could not squeeze through the opening. The Soviets had learned the lesson the hard way the year before but apparently The Cold War did not promote international communication and the Americans had to re-learn the same lesson, the same hard way, with the same death defying results.

In 2009 NASA astronauts were to make repairs to the Hubble space telescope which required them to leave their craft and perform a spacewalk. While out in space they would be tethered to the craft but that would be it. They could only hope nothing would go wrong because there was no back up plan and no nearby craft to save the five men. That was the exact story line in a recent Hollywood movie and something did go wrong but in the movie the astronauts sought refuge at the Intentional Space Station and were able to get there by using their jet packs. In reality the International Space Station is about 120 miles from the Hubble telescope which is too far for the jet packs to travel. Fortunately, the repairs went according to plan, there were no problems and no need to use those jet packs.

In 1997 a cargo space ship slammed into the Mir Space station and punched a hole in its side. It was thought that the Russian cosmonauts inside would be killed when the inside of the ship underwent a severe decompression. The cosmonauts had no tools to repair the damage so they made do with what they had, kitchen utensils. The two men managed not to panic and started sawing through the electric wires that connected the damaged module to the rest of their ship. Again they had to work fast because of the threat of decompression. They also had no space suits, which probably helped them work even faster. They cleared and then closed the hatch before outer space rushed in and killed all of them. You can only imagine the story they have to tell their grandchildren and that story just might have been made better, more colorful and more dangerous as embellishment seems to grow with the passage of time.

In 1985 there was a problem on board the Space Shuttle Discovery. The shuttle was supposed to release a satellite but the rocket would not ignite. Even NASA has problems from time to time with engines that won't start. To make a long story short, the astronauts went outside to take a look under the hood and see if they could get the satellite's rocket to fire. Problem was they were not supposed to do that because they had no training in those matters. One can only imagine how nervous they must have been tinkering around on that engine that just might go off at any time and take them with it. The rocket never went off and the two astronauts were able to get back inside alive, with another chapter in hand about their odysseys of space that they could share with their grandchildren. You might ask, why not just dump the satellite and the faulty rocket engine in space? Well they couldn't do that because they would contribute the mass of orbital debris already there, apparently some folks were aware of the problem before it became as serious as it is today.

In 1973 a pair of astronauts was outside of the Skylab space station when they had a problem. They were out there trying to fix a solar panel when the panel suddenly deployed and knocked the men right off the station. They were in the process of being hurled into the great unknown when their harness belts kicked in. The belts did not snap and kept them anchored to the station. One can only imagine the kind of whip lash they experienced, not to mention that split second of extreme uncertainly they had when they were not sure if their harnesses would snap.

In 1984 an astronaut put something called a Manned Maneuvering Unit on his back to perform a series of tests and maneuvers. The exercises took him in and around the shuttle's payload bay. The MMU was basically a jet pack that was not tethered to the space craft. The astronaut would travel 320 feet away from the shuttle Challenger which is a little longer than a football field and in the process he became the first astronaut to become a satellite that orbited Earth without any attachment or mechanical enclosure. If he had any problems with his jet pack he would not have been able to tell his children about the story.

In 2015 a European Space Agency astronaut encountered serious problems within his space suit. Apparently he almost choked to death by water that had somehow leaked from somewhere inside of the suit. The threat of drowning in space was real, water filled his ears and covered his nose. In the moments he had left, he managed to stay calm and started to feel his way around until he found the air lock and crew members were able to help him back inside. Once back inside he too had stories to tell his children.

Are we Looking in the Right Place?

It is hard to understand why intelligent life in the universe seems to be so rare when there are so many places to look. A large percentage of these planets are outside of our solar system and are called extra solar planets, planets outside our solar system that orbit a star. The search for alien life can easily become too large for practical purposes so let's restrict it to just our galaxy. Remember there are an estimated one hundred billion galaxies in the observable universe so to keep our figures manageable we only look at the Milky Way.

According to some of the latest research scientists say there are an estimated 100 million planets in the Milky Way that might be able to host some sort of life. About one fourth of those planets or maybe 25 million are orbiting stars that seem to be like our sun. New planets are always being discovered so this figure will likely soon change. For instance, one week in the spring of 2014, 715 new planets were discovered.

The criteria for habitable planets include: the planets must be the basic size of Earth and must also be about the same basic distance from their stars as Earth is from our Sun. If the planets are too close or too far away it would be difficult if not impossible for intelligent life to develop. The right distance or habitable zone from a star to a planet is about 90 million miles, we're 93 million miles from the sun.

If the planet is too small it might not have enough gravity to keep its atmosphere, like Mercury and to a lesser extent like many of the moons in the solar system including our own. If a planet is too big there is a good chance that it's gravity would be too strong and might crush intelligent life. That's because large planets usually have thick atmospheres made from hydrogen which often translates into tremendous surface pressures that would make life almost impossible. The ideal size for a planet to sustain life would be about one and a half times larger than Earth. So an ideal range for habitation might be no smaller than Earth and no larger than Neptune. Neptune is about 12 times larger than our planet. From this step we further refine our search to extra solar planets, again those that orbit stars outside of the solar system. At last check there are an estimated 2000 of these planets that we know of and probably many more that we don't.

What happened if life was so different from ours that it if we found it we might not even recognize it? Would that scenario increase the chances that intelligent alien life exists in the universe? That raises another question, what's the benefit of intelligent alien life so different from our own that we can neither recognize it nor benefit from it?

Searching for these extra solar planets is almost amazing as the planets themselves. Consider how hard it is to detect a much bigger and brighter star that is light years away. That star might seem like a little speck. Well the planet that it's orbiting is even a smaller spec, much dimmer and more obscure. In fact, they're almost impossible to see. So astronomers came up with a new search method. They apparently look at the stars the planet's orbit and then look closer to see what kind of effects the orbiting planets have on their host stars and that's the method behind this difficult search.

Extra solar planets orbit several type of stars, from ones like our own sun to red dwarfs and everything in between. Red dwarfs are of special interest for a number of reasons. They are the most common stars in the universe, so if life can exist around them then life might be relatively common in the universe. According to some of the most recent research about half of all red dwarfs are host to planets in their orbits. The sizes of the planets are promising because they are all from one to several times the size of Earth. Their numbers are also promising because there could be 100 billion red dwarfs in the Milky way. A few of these promising extra planets are listed below.

There is an extra solar planet called Tau Ceti, it is almost outside of its habitable zone and is about four to six times the size of Earth, might be too large for human habitation but it's gravity traps a lot of heat. It's about 12 light years from our Sun, its orbital distance falls somewhere between those of Venus and Mercury and it takes Tau Ceti about 168 days to orbit its star.

Gliese 581d is also promising and is close according to galactic standards, just some 20 light years away from our sun. It might be too big to be habitable, about seven times the size of Earth but despite its size it might be the first planet beyond Earth that could be capable of supporting oceans, clouds, rain and maybe even simple life. It might have an atmosphere with some sort of carbon dioxide mixture. Articles about the extra planet contained headlines like, 'Habitable Super-Earth" and '' An Exoplanet Fit for Humans." It orbits a red dwarf and has a sister planet called 581g.

Glisese 581g is that aforementioned sister planet and is about 20 light years away and could be two or three times as large as the Earth. Its year lasts one month meaning it orbits its star once every 30 days. It is one of the most Earth like planets found in the habitable zone but appears to be tidally locked; meaning the planet does not spin on its axis but instead shows its same face to its host star. To date, we have found no planet that is tidally locked that is capable of hosting life but Glisese 581g does have enough gravity to hold on to its atmosphere.

Tau Ceti e is also fairly close to Earth, about 12 light years away. It is about four times the size of Earth and might be too big for habitation. Its atmosphere could be too hot for advanced intelligent life but might be able to support simple life forms. Scientists estimate its surface temperature to be as high as 154 degrees Fahrenheit. Temperatures of above 140 degrees Fahrenheit are probably too hot as humans cannot cool ourselves off quickly enough to prevent heat stroke.

HD 85512B is a promising planet, 35 light years away and is about 3.5 times bigger than Earth and there could be water on HD 85512B. It orbits an orange dwarf star that is about three times the size of our sun. The surface temperature there is estimated to around 75 degrees Fahrenheit, suggesting it might have water but its gravity is about 40 percent stronger than Earth's which might be too strong to support life.

Kepler-22b holds some of the best promise for habitation. It is about 600 light years from Earth and is thought to be a little more than twice the size of our planet. Because its star is similar to our sun, Kepler-22b's surface temperature might be 72 degrees Fahrenheit. Its orbit is similar to Earth's, taking about 290 days to orbit its star and there might be water on its surface but scientists are not sure if that surface is solid or gaseous.

The Crown Jewel!

It can be easy to take Earth for granted considering its all that we've ever known but scientists say we should appreciate the planet because it might be the only one of its kind. According to one calculation, there's less than a.01 percent chance in a billion that another Earth is out there somewhere. Another calculation puts the odds even higher, truth is no one really knows but a growing consensus is Earth's so rare it might have been created by accident.

First there must a planet that's the right size, too small and there's no atmosphere, too big and its gravity might crush you. Then the planet must be the right distance from its star, too close and its too hot and vice versa. This is called the Habitable Zone which has already been mentioned and the size of the habitable zone is another factor. Habitable Zones change with the age and type of star, which might decrease the chances of intelligent life evolving somewhere out there. A star may become too hot, too soon before one of the planets has the chance to develop intelligent life, so the right star has to be stable for billions of years. Then you have to wait for the pot to stew, meaning planets are not immediately formed with life on their surfaces, these things take time, millions or billions of years. There's a strong chance that in its early stages a planet might have been super-hot where rock turned to liquid. It might have taken Earth a million years to cool off and in that time it might have been hit by asteroids, or another planet, which causes it to heat up again.

Just the opposite might be even worse, consider the possibility that if Earth was never hit by the asteroid that killed the dinosaurs 65 million years ago. Chances are we would not be here because the dinosaurs would still be here. So you get a better idea of how rare it is for something like Earth to develop.

Let's say you are fortunate to have the right planet and the right star and they're both the right distances from each other but you are in the wrong galaxy. There are thought to be at least 100 billion galaxies in the universe and most are not habitable. In fact, much of the universe cannot support life and is known as dead zones. Depending on with whom you speak as much as 70 percent of the galaxies can't support life which only further reduces the chances of finding a planet like Earth.

There also must be the proper arrangement of planets. Scientist say our solar system has the right configuration with the rocky planets up front and the larger gaseous ones in the back. Mercury, Venus, Earth and Mars are the inner planets and Jupiter, Saturn, Uranus and Neptune are the outer ones. According to theorists, the strong gravity of Jupiter sort of sweeps the area of asteroids which might have hit the inner planets with greater frequency and therefore caused a greater number of global extinction events in the past.

What are the chances of something like Theia happening? Remember Theia is the theory of how the moon was created. Without the moon chances are there would not be life as we know it on the planet. Prior to Theia, Earth spun on its axis much faster than it currently does meaning a day might have lasted eight hours. Those early conditions would have made it hard for life to develop. Then came Theia, the Mars size planet that hit Earth at just the right angle and just the right speed so as not to destroy the young planet and to slow is spinning speed to something more reasonable. Still more odds to overcome, immediately after the impact Earth wobbled too radically on its axis for complex life to evolve. Under those wobbly conditions ocean tides might have rushed in and out hundreds of miles, twice every day and winds might have been so strong that they would have scoured the surface clean of nearly everything capable of giving life. Debris from the Theia impact soon formed the Moon and Earth became stable.

Theia is also credited with saving the Earth's all important magnetic field and without that field Earth might have ended up like Venus. According to one of the more common theories, Venus was also hit with a similar sized planet as Theia but the results on Venus were devastating. The impact on Venus nearly stopped the planet from spinning on its axis. A planet must spin on its axis in order for the metal at its core to generate a magnetic field. The magnetic field repels the sun's harmful radiation. When Venus's magnetic field was destroyed the sun's radiation ripped away the essential parts of its atmosphere and that resulted in a runaway greenhouse effect on the planet. The end result, Venus has 800 degree temperatures and a zero possibility of supporting life.

So what are the odds that Earth was fortunate enough to have iron at its core? And what are the odds that when Earth was hit by Theia, Theia did not completely stop the Earth from spinning on its axis? You can see that at almost every evolutionary turn Earth has taken in its 4 billion plus year history it has beaten the odds for survival. There are still more odds for Earth to overcome. Consider the fact that it took about 800 million years for complex organisms to show up, we are fortunate that in that 800-million-year window Earth was not destroyed by a collision from another planet or a series of asteroids or was torn apart by a black hole.

Plate tectonics also developed by chance and they too are essential for life. Plate tectonics depend on a constant source of heat, like the one that comes from the center of Earth, they also produce a number of chemicals that are essential for life.

Another unique aspect of Earth is air and it's the only planet that has it. The air in our atmosphere has just the right composition for human habitation, about 78 percent nitrogen and about 20 percent oxygen. The atmosphere extends all the way down to the Earth's surface and rises several thousand miles up. It begins with the troposphere at the surface and rises 7 miles. Then there is the stratosphere about 7 miles high to 31 miles in elevation, followed by the mesosphere at 31 miles to 50 miles high and then there is the thermosphere from 50 to 440 miles and ending with the exosphere at 440 miles to about 6,200 miles.

Earth is also the only planet with an ozone layer which protects us from the harmful effects of the sun's solar radiation. At risk of over simplification, it basically forms a shield high up in the sky, beginning about 6 miles up and ending around 30 miles and absorbs most of the sun's harmful ultraviolet radiation. It we did not have this shield it is thought that the majority of humans would have skin cancer, maybe as high as 90 percent.

Many scientists say despite all that has been mentioned, it is water that makes Earth the special place that it is. One of the most basic of scientific beliefs: water is required for life. Other planets have water but it is not like the kind we have here. There are plenty of planets with frozen water, which we know as ice. There's a high probability other forms of water in the universe are too heavy to support human life. As mentioned earlier heavy water is about 10 percent denser than water here Earth. Heavy water has a different chemical composition than normal water. And the distribution of that normal water on Earth could not have been more perfect, not too much as to cover the higher elevations and not too little as to turn the surface into a desert. Earth has been able to retain its water long enough for humans to evolve where water on other planets has long since disappeared

Astrophysicist Greg Laughlin put a price tag on some of the planets in the solar system. Earth came in at 5 quadrillion dollars, Mars was valued at 16,00 dollars and Venus was nearly without value, worth only a penny. The value range between Earth and Venus is too large to really comprehend but suffice it to say that Earth is a jewel that has beaten the odds and we are the beneficiaries whether we appreciate it or not.

Risky Profession!

Harrison Schmitt is widely considered to be a very lucky astronaut. He flew on board Apollo 17 and was also the only scientist to walk on the moon. He was a geologist originally scheduled to be on board Apollo 18 but when that mission was canceled he was shifted to Apollo 17. If Apollo 17 was launched a few months earlier Mr. Schmitt and other crew mates might have been killed by a super strong solar storm. The flares were caused by magnetic energy in the sun's atmosphere and released tremendous amounts of radiation. Just one flare could be 10 million times stronger than a volcano and release billions of tons of radioactive material. During some of the stronger storms as many as 20 solar flares can be released in a day shooting out into space at about 1,000 miles per second. Some of these flames are ten times the size of Earth and what followed was one of the worst solar storms ever witnessed. If this solar storm occurred a few months later the result might have been disastrous. Apollo 17's crew might have been on the lunar surface collecting moon rocks with no protection other than their space suits. Remember the moon has no atmosphere or magnetic shield and therefore no way of deflecting or diluting the sun's radiation which would've been in full effect. Mr. Schmitt's avoided those dangers, collected lots of unusual rocks from the moon and later became US Senator from New Mexico.

The next astronauts might not have considered themselves lucky when they were struck by misfortunes but probably do now. Here are some of their stories. Think if your car lights went out just as you entered traffic, well that sort of happened to Space Shuttle Astronaut Chris Hadfield. He was ready to dock with the Mir Space Station when he had a system failure. His sensors stopped working just about 30 seconds before impact.

Mr. Hadfield had to maneuver the huge 250-thousand-pound shuttle to a small target on the Mir Space station that was about the size of a tennis ball. No time for stage fright because there was only about 30 seconds left. His next movements with the space craft would have to be perfect, if he hit too lightly he would bounce off, too hard and it would break apart Mir and might kill the three people inside. When he was about 30 feet away the sensors started telling him different things, at one point he was told he was 32 feet away and at another point he was told he was 20 feet away. In a situation like that you really just can't split the difference. One calculation could be wrong or both could be wrong and who do you ask in such a situation?

What happened next sounds like something out a Hollywood movie but it is documented fact. Because Mr. Hadfield knew the dimensions of the docking area, he put his thumb up to his eye and eyeballed it right through the shuttle window. That's right with all of the latest technology, he had to eye ball it with thumb up to window and likely with his other eye closed, just like an old time carpenter might have done. Turns out this astronaut had a dead eye because he was spot on, the target on the dock was hit dead center and with a few seconds to spare. After the feat he apparently said "we did it" maybe out of surprise or relief or maybe a combination of the two.

The story of Bob Curbeam reads like I novel, he might have wished it was but it was not. The NASA Astronaut was no stranger to space walks and was performing another one that he might have thought was going to be routine. It wasn't. He was on the outside of the International Space Station conducting upgrades when a high pressure line broke and covered him with fluid. Turns out the cooling line carried toxic ammonia which covered his suit. Problems often compound, this one did and now Mr. Curbeam had two. He couldn't go back inside because of the ammonia in the closed environment of the shuttle could be potentially fatal to the other astronauts. Mr. Curbeam also had no means to rid his suit of all the ammonia. Astronaut Curbeam realized that ammonia has a fairly low boiling point so he figured he would just vaporize or bake the toxic chemical off of his suit. The boiling point of ammonia is minus 28 degrees Fahrenheit. He did this by taking a bath, a solar one, that's right he baked himself in the sun for an extra 30 minutes and sure enough that did the trick, the ammonia was gone. So he took the world's most unusual sun bath, with no atmosphere or magnetic shield for protection, traveling at more than 15 thousand miles an hour, 100 miles above Earth, not to improve his hygiene but not save his life.

There's nothing like a fire to ruin your dinner and that's what happened to Astronaut Jerry Linenger. He was on the Mir Space Station when a fire broke out and threatened to spoil what was supposed to be the longest period an Astronaut spent in space. An uncontrollable fire on the Mir Space Station was called one of the worst case scenarios imaginable. The blaze lasted for 14 minutes but must have seemed like 14 hours. Making matters worse, the fire blocked their escape route to another space vehicle. The vehicle could only carry three people which meant they could only save three of the six-person crew. How to decide which three people would live and which three would die was going to be difficult. The fire was finally put out, no one was hurt and what might have been the biggest relief was no one had to make that tough decision.

Then there was the time a group of Korean astronauts had to ask nomadic herders for help. They also asked the herders if they might happen to have a cell phone so the astronauts could call for help. Seems the crew had some sort of malfunction during reentry which put them about 300 miles off course. When they landed on the ground crew members had to get out of the craft on their own. When they crawled out they just laid there on the ground, surely thankful that they were alive but in disbelief over what just happened and they survived. It was then that they were stumbled upon by nomadic herders in Kazakhstan who probably did not know what to think. The herders had to be persuaded that these strange looking beings were not aliens from space. Once they realized they were actually humans the herders helped to get the last crew member out of the craft and then they were asked if they had a cell phone which they did not. One of the crew members went back inside their craft and used a GPS device to call for help which arrived and brought to an end this tale that could have been the main story line for a big screen production.

Solar Storms !

Solar storms are caused when the sun ejects radio material into space that can disrupt electrical systems on Earth. The first solar storm recorded and apparently the largest ever since is called the Carrington Event, named after the man who discovered it, Amateur Astronomer Richard Carrington. The Carrington Event apparently was two storms in one, the first erupting at the end of August and the second at the beginning of September. The year was 1859. Most of the world's major newspapers published stories about its effects and the unusual aurora light activity over the north pole. Accounts recalled the lights as like nothing that was ever seen before which illuminated the entire night sky, using adjectives like brilliant and beautiful. From Baltimore, MD to New Orleans. LA to San Francisco, CA to London, ENG crowds of people gathered on street corners at night, certain they were witnessing a huge fire just over horizon but as the lights grew their concern turned to amazement with the realization no Earthly fire could be that large.

Some people thought that the spectacular light show must be coming from space. Other folks panicked and thought the light show was a sign from God or maybe an indication that the world was coming to an end. There were other reports that fire fighters rushed out to douse the flames only to realize after driving for 20 minutes that they were not getting any closer. One can only imagine their conversations on the way back to the fire station.

In the fall of 1882 astronomers discovered a huge spot on the sun's surface that spanned more than a couple thousand miles. Big sunspots are usually a good indication that Earth should prepare for the effects of a solar storm that could hit within a couple of days. Those effects would show themselves in the form of the bearings on compasses being off by almost two degrees. Telegraph communication was shut down in parts of the nation's mid-section. Signal bells on many railroads went off by themselves. Around dusk many people at different locations reported seeing what appeared to be a big green cigar like figure that was illuminated in the sky. Newspapers were inundated with all reports ranging from strange lights, to communication problems, to telegraph outages and Chicago's Stock Market reported problems for most of that day.

In the spring of 1921 there was an enormous sun spot discovered that spanned an estimated 94 thousand miles. That's so long that it could wrap around the Earth almost four times. The spot on the sun was so big that it could be seen with the naked eye if viewed with a protective lens. First it produced a magnetic bearing change of 3 degrees, then it impacted communication from the east coast to the Mississippi River. Then it knocked out the switching and signal system for the New York Central Railroad. Then it caused a fire at one of its control towers. It also caused a fire that destroyed the Central New England Railroad Station as well causing problems overseas where a fire destroyed a telephone station.

Then in 1940 came something called the Easter Sunday storm. There was never a good time for a solar storm but Easter Sunday could have been one of the worst. Millions of Easter Sunday telephone calls between family members were never connected. Western Union was hard hit with major problems at nearly all of its offices. A telephone cable between the United States and Canada had all of its wires tightly fused together from intense surges in voltages. The Associated Press had more than 180 thousand miles of wires and all of them were out of service. Almost every major company that dealt with telephone or telegraph communication scrambled to make repairs. Lines that were designed to carry less than 50 volts were fused together or blown apart by surges of electricity that carried 600 volts.

Then later in the fall of 1941 there was another solar storm that had worse timing than the one before on Easter Sunday. This one came during the playoffs of professional baseball between the Brooklyn Dodgers and the Pittsburgh Pirates. Full scale commercial television broadcasts did not begin until 1947 so many Americans were listening to the game on radio. The score was tied 0-0 when the sound went out for 15 minutes. For one quarter hour there were no reports about this important game. When sound did finally resume the score was four to nothing in favor of Pittsburgh. Brooklyn fans were hysterical. They pounded on one local station and demanded an answer. The one they got made them almost as angry as if they had no answer at all. They were told the sun caused the problem.

As mentioned earlier there was the solar storm of 1972 that erupted between the two shuttle flights, one being Apollo 17, had that storm erupted a few months earlier or later, it might have had fatal consequences for astronauts. On August 2, 1972 astronomers reported seeing three solar flares within a 15-hour period and the flares all had tremendous strengths. They were as powerful as anyone had ever seen and were almost at the limit of the scale used to categorize their strengths. After seeing that scientists issued a warning that Earth would feel the effects in about two days. Aurora's were seen from Nebraska to Colorado. Auroras are flashes of lights above the north pole and are caused when radioactive particles that include electrons and protons from the Sun slam into Earth's atmosphere and become active and their chemicals light up. The solar storm knocked out phone and cable services in places like Illinois, Iowa and other mid-western states. Canadian authorities reported strong electrical surges in their communication lines. The disruption in Earth's magnetic field was said to disorient pigeons who depend on that field for their sense of orientation. There were also reports that taxi cab drivers were getting requests from fares that came in from across the continents.

Since the Carrington Storm in 1859 there have been at least 100 solar storms to disrupt electrical and communication systems on Earth. Many usually hit within 18 hours after leaving the sun, including one of the most recent called the Halloween Storm. That happened in 2003 and knocked out a 450-million-dollar research satellite. Besides the inconvenience related to the disruption of communication services there were no reported cases in which a person was killed by a solar storm.

On Earth solar storms seem to pose no physical threat to people and here's why. Solar storms have been occurring for billions of years and we as humans, with the protection of our atmosphere, have simply adapted to and evolved with them as part of our natural environment. Apparently there was never a problem with a solar storm until humanity became advanced to the point where electricity became common and more specific electrical power became wide spread. It seems only after electricity was invented in 1879 did things like telegraph, radio and television communications start getting disrupted. A solar storm never knocked out a campfire.

A Recap of 2015!

The year 2015 left behind ample material for some of the strangest stories in space including the highest liquor laboratory known to humanity. A Japanese whiskey company sent samples of its wares to the International Space Station to see how these spirits would age in space. The company is called Suntory and the research on board the station will be conducted in an experiment module that is designed to find out more about the process that makes whiskey mellow. Identical samples were kept on the ground in Japan and a comparison would be conducted about a year later. The comparison will include a test taste but there is no word if the astronauts could participate.

Similar tests have already been conducted, one was with a specially designed coffee cup that should make it easier for crew members to drink in space and another was a martini glass that promises to be problem free in the weightless environment of outer space. The ultimate goal of the Japanese whiskey company is to create more mellow product. To the average whiskey consumer "mellow" means taking the burn out of the drink. The company is apparently prepared to go the lengths that are out of this world in its search for a more mellow product.

A scientific researcher made the bold prediction of what aliens from other worlds might look like and even went as far as to guess their average size. With the help of his calculations he estimated they would be the size of polar bears. His work included calculating the minimum size required for intelligent life and he based his measurements on the requirements needed to conserve energy on Earth. The end result was

650 pounds which is the average size of a polar bear. This work might reinforce the advice from another top scientist who warned humans not to seek out alien life because that alien life might dominate us. At more than three times the weight of the average human these polar bear sized aliens just might do that. The scientific researcher also suggested that if there is intelligent life in the universe the smaller animals probably outnumber the larger ones. His explanation was larger animals need more resources so if they reduce their numbers they are more likely to get those needs met. Also larger animals tend to live longer than the smaller ones, so the longer an animal lives the greater its chances of evolving to a more complex life form.

He went on to suggest intelligent life in the universe would have a better chance of existing on the larger planets. But other scientists disagree, stating one cannot accurately estimate the size of alien life based on life forms on Earth simply because we don't know enough about space and have yet to find anything resembling alien life in the universe. The critics also posit, the reason humans are so successful is not because of our size but the size of our brains, opposable thumbs and bipedal stances. Another critic pointed out that polar bears are threatened with extinction because of their inability to adapt to environmental changes and if they walked on two legs instead of four they would have a better chance of survival.

2015 was also the year scientists published their work after discovering a pair of pulsar stars, spinning extremely fast as they orbited each other. They are moving so fast that scientist believe they can actually warp the space around them. They've been called the "Dancing Duo" and they produce ripples in space known as gravitational waves. Scientists have measured these waves and have come up with a few predictions. One is, the duo will continue blasting through space at tremendous speeds and will continue moving closer together. The twins are called JO737-3039 and appear to be approaching each other at the rate of 7 millimeters a day. There are 25 millimeters in one inch so they're not in a hurry.

When the NASA probe Dawn flew past the dwarf planet Ceres it noticed something on the surface that could not be explained. The inexplicable feature is a white spot inside in a crater on top of a mountain. The mountain is thought to be about four miles high, making it as tall as Mt. Denali, which is the highest point in north America. Scientist don't know what the white spot is but they say it has some of the properties of salt. NASA likes the salt theory because it is believed there's an ocean beneath the surface of this dwarf planet. Ceres in the Asteroid Belt located between Jupiter and Mars.

The crater on Ceres was likely left behind from an asteroid or comet impact but there is still no explaining why that big mountain is just sitting there all by its self. The asteroid or comet impact could have blasted away some of the surface material and exposed the frozen ocean below. If that ocean has a composition that includes salt it could raise even more questions like might it be similar to Earth's ocean?

That year also brought something called the '' Kissing Stars" which are two massive stars that appear to be connected. They also appear to be the same size and instead of one star taking material from the other they could be merging together as one. The pair of stars are 160,000 light years from Earth and are thought to be the hottest, biggest binary star system ever discovered. They are called VFTS 352 and their heat and size might be explained by their proximity, located in a nebula that it thought to be home to the most active region of star formation in our galaxy. When the two stars touch they share more than 25 percent of their masses. This touching process is thought to generate tremendous heat. It takes about a day for the stars to orbit each other. The distance between their centers is thought to be about 12 million kilometers, not much by galactic standards nor by the standards of big stars. Together those two stars are thought to be 57 times larger than our sun and more than five times as hot. Many of the big stars don't fade away, they flame out and this could be the case here. One possible end would be the pair keeps on spinning until they merge into one big star and then blow up in one of the most spectacular light shows in the universe called a gamma ray burst. The other possibility would be something called a double black hole, which are the remains of a large binary system, or a system with two stars. Either way a spectacular ending to a spectacular life is predicted for the two kissing stars

Martian Tsunami!

A new report released in the spring on 2016 suggests water on Mars might have existed in large enough quantities to form an ocean and was impacted by an asteroid or comet that set off a giant tsunami. Scientists go on to speculate there might have been two tsunamis on Mars in the distant past maybe more than 3 billion years ago, each tsunami caused by an asteroid or comet that hit a few million years apart. The waves could have been four hundred tall and large enough to cover more than one American state. There's a high likelihood that the giant waves would have been reddish in color due to Mars's rusty rocks that are rich in iron oxide which is why it is sometimes called the Red Planet. One can only imagine such a ghastly site, a giant wall of red water several hundred feet high sweeping by at an amazing speed sand depositing boulders in areas they would not normally be found. The location of the boulders and other materials is part of what is called an up slow slope flow and is an indication of a tsunami. The huge rock and other debris were apparently swept up hill and left at higher elevations which could have only been done by the tremendous force of a large tsunami. Then the water ran back down hill and left behind more evidence in the form of channels there were cut into the Martian soil.

Apparently pictures and information suggest a large scale redistribution of rock material in the low land area on Mars's north side. It is an area where scientists have long thought might harbor an ocean some three or four million years ago when it was wet and warm. Today it is cold and dry. One problem with the long standing theory was the lack of a shore line but the new results suggest that there was indeed a shore line in the same area that would coincide with a large body of water.

As is often the case with new discoveries they can raise more questions and this one asks how would the size of Mars effect these tsunamis? Remember Mars is about half the size of Earth and gravity on Mars is about one third of Earth's. These facts help to explain why Mars has the tallest mountain the solar system in Olympus Mons. Olympus Mons is three times the size of Mt. Everest yet Mars is about half the size of Earth. Scientists say the weaker gravity and atmospheric pressure on Mars helped Olympus Mons to grow so large. So would have the tsunami on Mars been two or three times as large as a tsunami on Earth? This much is obvious, the more we know the more we need to learn.

Told You So!

Scientists recently discovered something called gravitational waves and have compared them to opening a window to the universe and seeing things that were never seen before. That window promises to yield an enormous amount of new and important scientific knowledge. One astrophysicist said the discovery will help chart new directions for observing the universe for the next century and maybe even beyond.

Seems these waves were proposed a 100 years ago but were never confirmed until the winter of 2016. There are no prizes for guessing who first proposed the theory. Gravitational waves are thought to be vibrations in space similar to those caused when a rock is thrown into a still pond. The vibrations in that pond are evidence of the energy released from the impact caused when the rock was thrown in the water which caused the waves to emanated outward. The same laws that apply to the rock in the pond scenario apply to many occurrences in the universe. Scientists say when a black hole collides with another one that impact sends ripples vibrating through space which are evidence of that collision. These vibrations can travel huge distances, vibrating for light years and when they reach us they deliver the cosmic information, it's evidence of a violent collision billions of years ago.

Problem is, these ripples or vibrations are hard to detect as they pass through people and objects on Earth. One reason is just like the ripples caused by the rock thrown into the pond, those ripples in space become weaker and smaller as they travel farther out from the center until they are so far away and so weak that they can't be detected. The same thing happened after two black holes collided more than a billion years ago and by the time their gravitational waves reached us they were so faint and weak they were barely detectable.

The machines that can detect these gravitational waves are almost as impressive as the waves themselves. The machines are housed in two large observatories that cost about a billion dollars and took some 40 years to complete. These machines are so sensitive they can pick up changes in wave lengths as small as one thousandth the diameter of a proton. A proton is one of the particles that make up an atom. 500 thousand atoms make up the width of a strand of human hair. The average human hand contains trillions of atoms, just a few examples meant to illustrate the sensitivity of the machine.

After decades of trial and error researchers had a major breakthrough after they used what has been called the world's most sensitive detection device they were able to identify movement that took place in less than one thousandth of a second. Researchers say at first, these black holes circled one another about 30 times a second and then sped up to about 250 times a second until they finally collided in a big, violent type of fusion like unification.

Imagine swinging a ball at the end of a rope as fast as you can, the maximum rotation would be about two of three times a second. Now imagine a black hole, more than a million times larger than Earth rotating at 30 times a second and accelerating to 250 times a second, the speed, energy and distance covered in that one second are so mind boggling that it really cannot be fully comprehended. So try to imagine the Moon spinning around the Earth 30 times a second.

The gravitational waves are also hard to detect because of something called something called general relativity. The theory of general relativity was proposed by the same guy who predicted these gravitational waves. The theory of general relativity is extremely complex but for our purposes it has two branches, the speed of light is constant for all people or objects and the laws of physics are the same for all people or objects that are moving at relative speeds.

These black holes were massive, one thought to be 35 times the mass of our sun and the other was a little smaller, both collided with each other about 1.3 billion years ago. They were also extremely fast, traveling around the speed of light, 186 thousand miles per second and covering six trillion miles a year.

Again the relativity theory, which is about 100 years old suggested that once these waves were set in motion their speed would be constant and at some point their energy would be displayed with the help of gravity. And that's what scientists heard and recorded in 2016. Scientists say as the two black holes were just churning away out there in the far reaches of the universe for billions of years, sending out ripple after ripple which were interpreted as gravitational waves. After billions of years and who knows how many miles those gravitational waves, now reduced in size and strength by time and distance, finally washed ashore on Earth where there was that super sensitive machine that was able to pick up the faint vibrations.

So what does this discovery mean? Some experts compare it to when the structure of DNA was determined, others say it is on the same level of the Higgs boson particle which is the particle thought to give mass to all matter, while others say the discovery will produce a Nobel Prize. For science it means that there is a new window open on the universe that was previously closed. Apparently it is not a matter of if new scientific knowledge will be discovered but instead when and how much. It is the first detection of gravitational waves, the first detection of black holes, and it was the first proof that under certain conditions huge, fast moving objects in space create vibration waves that move out at the speed of light and that's just for starters. It's a completely new way of looking at the universe which can be good and bad. Good because our knowledge of the universe is expected to grow by leaps and bounds; bad because some of the universal laws that were thought to be established could very well be disproved which would mean we didn't know as much as we thought.

Life on a Magnet!

Earth has been described as one giant magnet, which is the short explanation of why it has a protective magnetic field. A longer explanation of the origin of the Earth's magnetic field is a bit more informative. For starters we should probably take a look at atoms, some of the smallest building blocks of life. Everything we can think of is made up of atoms, from hair and skin, to books and cars and everything in between. They have been called little collections of electricity. Atoms have electrons which are appropriately named and are charged. Most of the time these electrically charged atoms work in concert and move in the same direction which creates a magnetic field or magnetism. The direction of their spin determines the direction of the magnetic field.

In other words, magnetism is caused by the motion of electric charges and it works most efficiently with certain metals. The Earth's inner and outer cores are made of metal, the outer core is liquid and intense internal pressure keeps the inner core solid. The outer core moves and when it does it produces electricity and the magnetic field.

Here's how our magnetic field protect us from solar storms. When the Sun's throws off its harmful electrically charged particles, Earth's magnetic field repels them back out into space just like two magnets on a kitchen table push each other way. Magnets only work with certain kinds of metals and don't respond to things like glass or plastic, or even materials like gold, copper silver, aluminum and the likes. Magnets work with iron and nickel which are abundant in Earth's core and gives our planet it's magnetic field. Magnets either push or pull other objects that are charged with the right metal compositions. The magnetic field of Earth stretches out for some 37,000 miles into space and is called the magnetosphere, appropriately enough. Solar winds are charged particles from the Sun constantly sweeping by Earth and when they slam into the magnetosphere air molecules heat up and glow. Occasionally an unusually strong blast of these particles breaches Earth's magnetosphere and they rush in toward the planet. But Earth has a second line of defense in its inner radiation field which directs the deadly particles toward the planet's polls where they are dissipated. Red and green colors indicate oxygen and red and blue indicate nitrogen which are known as the auroras. So when the lights are on so is Earth's protective security system and without it our planet would be as barren and lifeless as Mars. Particles from the sun can be just like x-rays to the human body and in large does can break down our DNA and cause cancer. In short the Sun's harmful particles don't directly hit Earth but instead are forced around it, by its magnetism.

Other particles that are even more dangerous don't even make it inside our atmosphere are collected outside in places called the radiations belts. It's well understood that these belts are dangerous areas for both man and machine. It is also strongly advised that any craft from Earth should move through these belts as fast as possible and only with extremely strong shields. The belts were discovered in 1958 by James Van Allen and are called Van Allen Belts. One such belt is in the magnetosphere and the other is in the plasma-sphere. The plasma-sphere is actually inside of the magnetosphere.

The plasma-sphere is fairly stable but the magnetosphere is not and can expand and contract. The magnetosphere can expand all the way to the International Space Station and to other satellites as well. When the belts were first discovered they were so strong they were mistaken for a nuclear test. The charged particles in these belts are made up of plasma, which surrounds the sun and fills the universe. They're also invisible to the naked eye. Many of these particles move at the speed of light and can threaten to disrupt some of the several hundred satellites that we use for communication and navigation. The most intense area of these belts is as high as 12,000 miles above Earth. It is also thought that the single biggest obstacle to a trip to Mars would be for astronauts and their equipment to overcome the prolonged exposure to solar storms and their intense radiation.

Change is Coming!

If everything changes then get ready for something different in the direction of Earth's magnetic field. In this case change doesn't happen very often, according to scientist about once every 200,000 thousand years. The last time Earth's magnetic field change direction was almost 800,000 years ago. So we might be 600,000 years overdue.

It's thought that Earth's magnetic field is generated deep within its core. Iron and nickel are found there in abundance and they churn in a liquid state that creates convection. Convection is the movement within a fluid where the hotter, less dense material rises and the cooler, denser material sinks. This churning convection generates our magnetic field and the magnetic field protects us from the sun's harmful particles.

It is also believed the process is sustained by something called the Geo-dynamo. It is not completely understood but is thought to contain electrical currents that produce convection and involves the spinning of at least two liquid metals, nickel and iron in creating the magnetic field. But there is really no way to confirm the Geo-dynamo because it is located 4000 miles into the Earth's core, it is a huge reservoir of highly compressed liquid metal weighting trillions of tons. Scientists also believe that the Geo-dynamo creates new magnetic fields that usually move in the same directions as the previous ones but sometimes new magnetic fields are created in the opposite direction due to some sort of disturbance that is not completely understood. The directional changes are called instabilities. Smaller instabilities are fairly common and usually don't last long. Some of the larger instabilities behave differently and can last much longer. When this happens the new polarity takes over the entire core which forces the liquid material into the opposite direction. This is the current understanding of how the poles are reversed.

So what will happen when this pole reversal occurs? Some folks predict it might be violent shock similar to the football player running in one direction and then getting hit by a linebacker which instantly sends the player flying in the other direction. Under this scenario anything not anchored down tightly to the Earth would suddenly be sent flying up in the air, which would include people, furniture, cars and whatever else was sitting on the surface. Scientist say that won't happen. Another theory of what might lie ahead when this overdue pole reversal comes calling is the end of the world as we know it. North would become south and south, north. The continents would suddenly get launched in a different direction. Tremendous earth quakes would break out. Major cities would be engulfed by massive fires when utility and electrical lines were ripped apart by the sudden directional change. Many species would become extinct and the climate would change for the worse. Scientist say not so.

Pole reversals have happened in the past and there doesn't seem to be any evidence in the geologic record that it caused mass extinctions or even anything significant to the Earth. Animals like homing pigeons that use the magnetic field for navigation might be disoriented, but that seems to be about it. During this time there won't be a north or south pole which might impact computers and some navigational systems but updates and modifications to software and GPS systems should not be adversely effected.

Evidence of reversals are left behind in rock formations but they don't explain how it happened. The geologic records show these pole reversals have happened in the past and there is evidence to suggest another one could beginning soon. Pole reversals take a long time, maybe 10 thousand years. So it's a slow process during which the magnetic shield becomes so weak that more poles appear in addition to the ones in the north and south and after that, the process would start to build more strength and realign itself in the other direction.

Problem with a weak shield is exactly that, it's weak. Scientists don't know how long the shield would remain weak while the pole reversal process runs its course. If it takes up to 10 thousand years, the shield could be weak for half that time. That could mean Earth's magnetic shield would be weak for 5 thousand years meaning humans might be more exposed in that time to the harmful effects of cancer causing particles from the Sun. So when the poles change, the number of melanoma and other and skin cancer cases could significantly increase and the process might have already begun. A weak magnetic shield could lead to the formation of ozone holes. There is evidence to suggest that a hole in the ozone layer would allow more of the Sun's ultra violet radiation into the Earth's atmosphere which would increase the cases of cancer in general and skin cancer in particular. Simply put, the ozone layer is a protective shield of gas in the atmosphere. It is thought that if it is reduced by one percent that would translate into a five percent increase in the number of skin cancer cases among humans.

There is even one theory that posits the demise of Neanderthals, which were our previous evolutionary cousins, happened because of one or more ozone holes created a weak magnetic shield. But according to scientists the pole reversal was too weak to bring about a total reversal and somewhere along the line the process was stopped but maybe not before the Neanderthals became extinct.

The next time the poles might reverse could be soon than later, again we could already be some 600 thousand years overdue. Scientists say there is strong evidence that the strength of Earth's magnetic field has been decreasing at a pretty alarming rate for the last 160 years. Could this be clear evidence that the next pole reversal has already started?

The sun seems to follow a predictable two cycle period of activity every 11 years or so. The solar maximum and the solar minimum. The maximum, just as the term expresses is when activity on the sun's surface is at its maximum and the minimum is just the opposite. Sunspots during the maximum period can last for weeks and appear much larger, they are dark cooler areas on the Sun and can be more than twice the size of Earth. It is thought that sun spots are caused by tremendous magnetic activity that coincide with the release of solar flares and coronal mass ejections. Flares and coronal mass ejections are basically explosions on the sun's surface that release unimaginable amounts of energy. That energy includes radiation, heat, light and plasma that are shot out into space at more than a million miles an hour. All of these problems could become much worse it they occur when Earth's magnetic shield is weakened. Sun spots are also not dark but are really about as bright as a full Moon but because the Sun is so bright the spots appear to be dark against the Sun's blinding background. During minimum periods the spots are small and don't last for long. The most recent cycle is Solar Cycle 23 which began in 2000 and ended in 2009. Most cycles last between nine and fourteen years. The current cycle, Cycle 24 started in 2008 and began a period of minimum activity which could be the smallest cycle in more than a hundred years. The solar cycle has profound effects on Earth's magnetic shield and its ability to protect our planet's vast electrical grid from crippling power surges like the one in 1989 that knocked out power for nine hours to millions of people in Canada. It is commonly thought that the maximum periods of solar activity pose the greatest risks to Earth's electrical and communication grids due to the increased activity in solar flares and coronal mass ejections. The reverse is thought to be true during times of minimum solar activity. And as a reminder, all of these problems could be made worse if a solar cycle occurs at the same time Earth's magnetic shield is weak.

Where Did They Come From?

Science has produced a number of theories regarding where stars, planets and even the universe came from but seems to be perplexed when it comes to galaxies. Stars and planets make up solar systems, and solar systems make up galaxies and galaxies comprise the universe but it still does not explain the origin of galaxies.

Galaxies don't seem to be scattered at random throughout the universe but instead seem to be placed in clusters or groups. One theory of galactic formation suggests that they were created from the gas left over from the Big Bang. Another theory posits that gas from the Big Bang created stars and planets all over the universe and they later migrated or gravitated toward galaxies. Both are similar and neither has been completed accepted.

The leading theory puts forth that galaxies were formed very early in the universe, maybe one hundred million years after the Big Bang. The first galaxies might have been formed from huge concentrations of dark matter. Dark matter cannot be seen, does not put off light nor energy and is thought to make up a large portion of the universe. It is also thought that dark matter does not contain the same elements as regular matter such as protons, neutrons and electrons. Scientists are not sure what makes up dark matter and are not even certain it exists. But they think it does exist by the gravitation influence that dark matter puts on objects that are nearby, or put another way dark matter might hold the stars and galaxies in place. This dark matter pulled together huge amounts of atoms and under intense gravity were forced together under enough pressure to create fusion. Much the same way that stars are born. These galaxies grew in size which gave them the requisite gravity to attract smaller galaxies which meant the bigger galaxies became even larger.

One of the most easily understood theories is after the Big Bang, the universe was full of energy that turned into atoms and after enough time gravity brought the atoms together in large numbers where they formed into galaxies, that contained solar systems, which contained stars and planets.

Then there's the question of other universes, touched upon earlier in the section about multiverses. The theory of multiverses states that there are an infinite number of universes out there. There are no known laws of physics that would support such a theory but there are no laws that disprove it either. We are not going to be able to travel those great distances in outer space to observe for ourselves but we can put our hope in future generations of space telescopes to provide an answer.

We've Sprung a Leak!

As if the galaxy was not mysterious enough, add into the mix something called the Fermi Bubbles.

Our galaxy, the Milky Way is a spiral galaxy with three components, the disc which contains spiral arms, the halo and the nucleus or the central bulge and all of it spans 110 thousand light years across. Lay the disc on its side and it looks like a flat and level plain but in its middle and extending out from the top is a huge bubble and there is another bubble extending out from the bottom. The bubbles shoot out a distance that seems almost as high as the Milky Way is long, maybe more than a thousand times the mass of the sun. They are so large that under the right conditions they would take up most of the night sky, at certain times you might look up at see them but not know it because the human eye cannot detect gamma rays or x-rays so they remain invisible to us.

A closer look seems to indicate that gas is rushing out of our galaxy at a rate of two million miles an hour and is thought to create these massive bubbles. Scientist have no idea what is causing this huge gas leak or the consequences it might have for the galaxy or Earth but they have discovered that gas from one of the bubbles is racing toward our planet. The gas from both bubbles seems to have been racing outward for about two million years, which was about the time our human ancestors took their first steps upright on two feet.

Scientists also think the gas shooting out from the bubbles could be the remains of earlier stars. Stars populate the universe in sets of generations, the first generation stars contained just the simple elements like hydrogen and helium, which were too crude to help form life. When those first stars blew up they seeded the second generation of stars with heavier elements. So the gas shooting out from these bubbles could be one of the natural mechanisms by which stars are seeded and become more complex with heavier materials. The bubbles were only discovered in 2010 so scientists need more time to figure out if this theory has any validity.

Coming out of these bubbles is the most powerful energy known, something called gamma rays and they were first discovered by using the Fermi Gamma Ray Space Telescope which is how the bubbles got their name. Another point of view suggests that the bubbles are actually in the gamma rays. Astronomers have used the Hubble Space Telescope to look deeper into the galaxy and found a quasar was behind the bubbles. A quasar is a quasi-stellar radio source and have been compared to a black hole feeding. When material like part of the galaxy gets too close to the black hole it forms an accretion disk and under intense gravity becomes denser and super-heated to the point where it ejects tremendous amounts of radiation in one of the brightest light shows ever seen in the universe. It is also thought that quasars represent some sort of stage in the evolution of galaxies. So could the bubbles mean part of our galaxy is in danger of being consumed by a black hole? Scientists don't think so.

So are the Fermi Bubbles reshaping our galaxy? The galaxy has been around for 13.8 billion years and it's thought that the bubbles are only two million years old. Scientists don't know if this new figure 8 shape will become the permanent configuration of our galaxy. Keep in mind the Milky way could be as large as 180,000 light years across, so who knows what would happen if a structure that big changed permanently.

They also don't know what the long term effects will be of gas rushing out of the galaxy for the last two million years at the velocity of two million miles an hour. Like many theories of the universe what expands must at some point also contract. If these massive bubbles do contract back into the galaxy what will that mean? Either way the galaxy seems to be changing, which should not come as a surprise. Everything in the universe changes, all of it set in motion millions and billions of years ago. Scientists scrambled to come up with explanations for this behavior. One theory is something must have caused the gigantic eruption like possibly a large cloud of gas. The gas cloud could have collapsed over the center of the Milky Way, which is thought to be occupied by a black hole. As the cloud moved closer to the black hole that cloud became super-heated to the point where it might have exploded and in the process blew the surrounding gas far out into space and created the Fermi Bubbles. Another theory that is less popular could be the released gas caused by the Fermi Bubbles is too small to be concerned with when considered the Milky Way is 110,00 light years across as gas rushing out at the rate of two million miles an hour constitutes nothing more than a minute leak in a structure so large, again a less popular theory possibly because it is a boring.

Don't Blame the Messenger!

The NASA probe Messenger taught us a lot about the planet that is the least explored. Probably the main reason why we know so little about Mercury is it's so close to the Sun. Some astronomers say Mercury is too close to the Sun for us to learn much, the planet is about 36 million miles from the star. The temperatures on Mercury can be as high as 800 degrees Fahrenheit in the light and almost 300 below zero in the shade. Besides its close proximity to the Sun, another reason for the extreme temperature difference is Mercury has no atmosphere to retain its heat at night.

Messenger taught us about Mercury's average density, second only to Earth's in the Solar System. It has the thinnest atmosphere. It is the smallest planet in the Solar System and has a magnetic field that stretches across the entire planet. We would probably know much more about Mercury but there has only been one prior probe to visit it and that was the Mariner probe in the mid-1970s. Mariner was the first probe to utilize the gravity assist method of space travel. So scientist have known how to sling shot space craft around planets for almost 50 years.

Messenger was launched in 2004 at a cost of about 46 million dollars and was designed to tell us more about Mercury's composition, overall environment, its magnetic shield and its extreme temperature change. The probe was built for the long, slow haul but is also capable of great bursts of speed. Messenger's endurance was illustrated by the fact it traveled almost 5 billion miles which included 15 orbits around the sun. One reason for Mercury's long trip was the Sun: probes and other space craft that land on planets can perform a technique called aero braking. That's when the probe sort of skids through the planet's atmosphere to reduce its speed. Problem is Mercury doesn't have an atmosphere. So if a probe tried the aero braking maneuver on Mercury it might get pulled into the Sun because of the Sun's strong gravity and close proximity. So Messenger dumped its speed by flying around Earth, Venus and Mars and letting those planet's gravities slow it down. Sounds like aero braking could be compared to a gravity assist in reverse?

Messenger's average speed was a mind boggling 84,000 miles an hour which is more than twice the speed of the many asteroids that are tracked, or about five times the speed of space shuttles traveling in Earth's low orbit. Remember the asteroid that exploded over Siberia in 2013 was traveling 40,000 miles an hour. At other times Messenger sped up to more than 140,000 miles an hour. There was no explanation for the increase in speed but it certainly was impressive.

When Messenger was launched fully loaded it weighed in a 2,400 pounds and about half of that bulk was fuel. Apparently it needed all that fuel to slow down enough to enter Mercury's orbit and not smash into its scorching hot surface. Messenger is the size of an average home entertainment center, 4.7 feet tall and 6.0 feet wide and 4 feet deep. So it's small and heavy, less than five feet tall and weighing more than a ton. The probe needed extra protection from the sun so it was equipped with a special parasol. A parasol could be compared to an umbrella or shield. The parasol was heat resistant, stands about 8 feet tall and was coated with a highly reflective substance. The front part of the parasol can get as hot as 700 degrees, comparable to the inside of an oven when it is on self-clean. Remember the Sun on Mercury is thought to be more than 10 times as bright as it is on Earth and maybe 10 times as hot. The back side of the parasol stays a cozy 70 degrees when Mercury's temperature in the shade can be several hundred degrees below zero.

Despite the years of preparation, including the years needed to travel to its destination, Messenger's mission lasted just two days. That is two days on Mercury which is equal to 176 days on Earth. Remember a day on Mercury lasts longer than its year because Mercury orbits the Sun faster than it rotates on its axis.

Messenger's good byes are final because it was sent on one a one-way trip. When all is said and done Messenger won't have enough fuel to make it back home to Earth. Messenger was launched in the summer of 2004, successfully completed its main mission in 2012 and crashed in to Mercury in the spring of 2015. Messenger's impact added another crater to Mercury's surface but this impact did not come from an asteroid but rather from a machine sent by humans, arguably just as amazing as any of the other facts from space exploration.

Head in the Clouds!

No one ever accused NASA of dreaming small. The space administration has plans to construct a floating space station over Venus. Remember Venus could not be more inhospitable, with an average temperature of nearly 800 degrees Fahrenheit, as hot as an oven set on self-clean and clouds that rain sulfuric acid like the acid in a car battery. Its surface pressure would likely crush humans because it is more than 90 times greater than Earth's. Only one probe sent from Earth lasted long enough in the intense heat of the Venetian atmosphere to take pictures before its circuits melted, all other craft sent by humans were destroyed by searing temperatures before they could record images but yet NASA wants to build a floating space station over the oven like planet.

Apparently conditions are a lot better just a little way up. According to NASA just a mere 30 miles above the Venetian clouds is another world, one that has the potential to be quite similar to Earth's. The pressure in the atmosphere is similar. The gravity is similar, thought to be a little weaker than Earth's. Similar gravity has its benefits, allowing astronauts to stay in the Venetian space station longer and decreasing health problems, like bone loss and muscle atrophy that crop up in a zero gravity environment. Of course the station would be hot, probably just barely tolerable but the heat might be managed with a cooling system. NASA also reports that the altitude of a Venetian space station would be sufficient enough to protect its occupants from the Sun's solar radiation.

Venus is closer to Earth than is Mars. Venus is about 25 million miles away or about a 100 times farther than the Moon is from Earth. Mars is little more than twice as far at 54.6 million miles out. It could take at least three months to travel to Venus. Of course the trip would have to be timed just right and coincide with the planet's nearest approach during their orbits.

The trip might begin with a probe that would venture down to the Venetian surface and hopefully not burn up in the process. Data would then be sent back to the Venetian space station to prepare for the next stage which would include deployment. NASA is not quite sure how many crew members would be deployed on the first mission, but it might last about 30 days. That would be followed by groups of astronauts, possibly in pairs that could spend up to a year in the Venetian space station. The goal would eventually be to have the astronauts establish a permanent colony up there some 30 miles above the Venetian clouds. While its head may be in the clouds the focus of the Venetian space station is firmly on the ground as NASA plans to use technology that already exists or is close to existence for this mission. The plans are for the not so distant future, maybe 10 to 20 years away and could suggest another approach to space exploration.

Because of the present technological limitations, it's not likely humans will to travel to and even colonize nearby planets anytime soon. So are plans for a Venetian space station the next best thing? As one scientist put it, a Venetian Space Station would give us sort of a "head start" or a "practice trial run" for the loftier goals of actually colonizing a planet. Those who advocate building the Venetian Space Station say its gravity can be compared to Earth's and would be easier to reach than Mars. They also say next world humans might encounter could be the one we build over Venus.

Does Proximity Equal Similarity?

The closest system of stars to our galaxy is called Alpha Centauri. It consists of three stars which could be called a, b and c, for Alpha Centauri A, for Alpha Centauri B, and Alpha Centauri C. A and B are part of a pair and C stands alone. A and B are sizeable stars comparable to our sun. The system is about 4.5 light years away and has been called the most famous star system in the sky. Because it is so close scientists are trying to determine if the Alpha Centauri system is similar to our solar system.

A new planet was discovered orbiting Alpha Centauri B, scientists automatically wanted to know if it could support life, which it could not. It is about the right size for a habitable planet but it's not in the habitable zone and its far too close to its star to support any semblance of life. The new planet is called Alpha Centauri Bb, apparently the lower case b denotes a planet. It takes about three days to orbit its star, or put another way, a year on Alpha Centauri Bb takes about 72 hours. The planet is too close to its star, just about 4 million miles away. Remember Earth is about 93 million miles from the Sun, so scientists estimate the surface temperature of Alpha Centauri Bb to be about 2200 degrees Fahrenheit, far hotter than what is required to melt rock.

Scientists are optimistic there might be another planet lurking in the shadows of the triple star system, a planet that might be in the habitable zone, farther away from one of the stars. Scientists use computer models to help in the search. Their research suggests there could be one or more planets in the system that have extremely large orbits which could be one reason why the planets have not been seen through telescopes. Apparently two of the three stars in Alpha Centauri are similar to our Sun in size, luminosity and age which again is cause for optimism that another Earth like planet is hidden somewhere in the system. Results from one computer simulation suggested that a planet might have been able to form on the out skirts of one of the major stars which would put that planet squarely in the middle of the habitable zone.

Scientists were not discouraged by the fact that there is likely no life on the planet orbiting Alpha Centauri B and have begun to speculate the possibility of sending a probe to the system because of its close proximity. 4.5 light years translates into about 25 trillion miles, much too far for any human mission to be considered. But what about a probe that has already been dispatched on a different mission? Could that probe be re-programed at the end of one mission in order to begin another one?

All of the questions and scenarios apply to the New Horizons probe. New Horizons was launched in 2006 and flew by Pluto in 2015. Pluto is about 4.5 billion miles from Earth so if by some technological marvel New Horizons could be modified for a new mission toward Alpha Centauri it still might take tens of thousands of years to get there. New Horizons is one of the fastest machines ever built but the distances are just too great to entertain any optimism. New Horizons travels at more than 36,000 miles an hour about 100 times faster than a jet and at that speed it would take about 75,000 years to reach Alpha Centauri. But scientists say if and when the time comes that humans can travel great distances in space Alpha Centauri will likely be the first place we visit.

While it might be nearly impossible to redirect New Horizons, it would be even more daunting to reconfigure its predecessor Voyager. Voyager was launched in 1976 and after some 35 years in space it left our solar system. That feat in itself is amazing. The probe has traveled farther from Earth than any machine ever built. After all that time and all those miles Voyager is still sending data back to Earth. That's a stunning feat when you consider Voyager is father from the Sun than is Pluto. It is more than 12 billion miles out and gets fathers all the time. A signal from Voyager travels at the speed of light and takes about 17 hours to reach Earth. At last check Voyager was traveling around 38,000 miles an hour and is expected to enter the Oort Cloud in about 260 years. The Oort Cloud is the area just outside of the solar system and is about 93 trillion miles away. Remember Alpha Centauri is only 25 trillion miles away so the idea of redirecting the probe is not entirely unreasonable to consider. It is headed to a star called AC + 79 3888 which is more than 17 light years away. There is no indication that it will reach its destination and is expected to stop working sometime around 2025 when its thermoelectric generators run out of power. Despite the great distance Voyager can still receive commands.

The Alpha Centauri system can be seen with the naked eye. But because it is so far away all three stars appear to be one bright star. Alpha Centauri has been called our "next door neighbor" and has been seen on Earth since antiquities by the early Egyptians. It is the closest and brightest star system that has an extra solar planet and holds a special place in our cosmic ambitions due to the fact that its has been called our neighbor. Question is, will we ever get to know our neighbor?

Help Wanted!

Apparently NASA needs help in tracking asteroids which could be interpreted as help wanted sign that might be of concern to all of humanity. One of NASA's efforts to locate asteroids is called the Near Earth Object program which was created in 1980s. The goal is simple, to search, identify and catalog objects like asteroids, comets and other similar material in space that comes within 28 million miles of Earth.

According to some researchers including a group called B612, there were about 26 meteorites that struck Earth in the last 15 years or so and all of them were big enough to level a major city. We'll learn more about the B612 Foundation later. The good thing is all of these meteorites struck in areas that were not heavily populated like the ocean, water covers about 71 percent of Earth's surface.

Case in point was the asteroid that impacted Earth in the winter of 2016. One headline read "NASA Says a Fireball Crashed into The Atlantic, But No One Noticed", and apparently no one did. According to NASA it landed in the middle of the ocean at 1:55 pm on February 6, 2016. The report described it as an exceptionally bright meteor that exploded in the sky well above the water. Compared to other meteors this one was fairly small but was still big enough to produce the equivalent of about 13,000 tons of TNT which was more than enough to level parts of populated areas.

Talking about an unannounced close call, an asteroid the size of a house was discovered just a few days before it made its closest approach. It was discovered in the fall of 2014 and if it were on a collision course we couldn't have done anything about it. Again at the size of a house, about 60 feet across and traveling in excess of 20,000 miles an hour it posed a real threat. Had it hit Earth it could have killed millions of people. It was definitely in the local neighborhood of the solar system, about 21,000 thousand miles away, actually flying under some satellites. No word if this asteroid came from that blind side, mentioned earlier as the direction of the sun in which telescopes cannot look. We do know that when the asteroid was nearest Earth it was too dim to see with the naked eye. So if worrying about the blind side was not enough, now there's concern over asteroids too dim to see until they are nearly upon us.

Problem with tracking these asteroids and meteoroids is there are not enough people tracking them and too few resources to help with the effort. According to a report published in 2014 the NEO program needs to be better organized, managed more efficiently and equipped with a bigger staff. These reported deficiencies mean the program will not meet one of its main goals before its self-imposed deadline. For about the last decade scientists with the program have been tracking asteroids that are bigger than about 450 feet. The goal was to catalog most of them by 2020. At last check there were more than 12,000 asteroids larger than 450 feet and less than 15 percent of them have been tracked with their orbits were plotted, which means that 2020 deadline might not be met.

There are apparently almost a thousand NEOs that are larger than a thousand yards and about 90 percent of them have been discovered. But there are many smaller ones that go undetected. Don't forget an asteroid about 35 yards, traveling more than 15,000 miles an hour could kill millions of people. Comets seem to be rarer but are every bit as dangerous. At last check there are about 96 comet NEOs that are tracked. Again these things hit us all the time but most fall in the ocean or remote areas.

According to some researchers Earth is hit on the average of once a year by a comet or asteroid that is larger than 12 feet. And around every five years Earth is hit with a significantly sized asteroid that has enough energy to equal the Hiroshima atomic bomb. And about every 3,000 years Earth is hit with something much larger, like the asteroid that exploded over Tunguska in 1908. That asteroid never made it to the ground but still leveled some 770 miles of forest. Had it hit today in or near populated areas, millions of people might have been killed which has scientists pronouncing were are living on borrowed time, they just don't know how much is left.

There are also been plenty of near misses, too many to document but here are a few. In 1972 Earth was grazed by an asteroid that came within 34 miles of striking its surface. Put into context on just how close that was, Earth's circumference is little more than 24 thousand miles and this asteroid came within a mere 34 miles. It was called the Great Daylight Fireball and was seen over the Rocky Mountains from the Southwest to southern parts of Canada. Then 17 years later came another one but this time it did not come as close. This asteroid was about a thousand feet in diameter and flew by some 430,000 miles away. Had it hit it might have caused the biggest explosion followed by the most damage and injury in history.

The following year Earth was grazed by another asteroid, this one came within 60 miles of the surface. There were many close calls in between, but ten years later another NEO came calling, this one came as close as 76 thousand miles and if it hit Earth many of us probably would have been goners. That asteroid would have exploded with a force of nearly 100 times greater than the Hiroshima bomb. And then there is the asteroid called 2013 TX68 scheduled to fly by in the fall of 2017. It flew by a few years ago but was much farther out, more than a million miles away. This asteroid is small by galactic standards, measuring only 100 feet and could come as close as 11,000 miles from Earth. Of course when it is that close, you wonder what are the chances of impact? They're thought to be about 1 in 250 million.

Some of the people in the trenches tracking asteroids work at NASA's Jet Propulsion Laboratory and when it comes to searching for space rocks they say it all comes down to random chance. Asteroid trackers openly admit there are millions of asteroids and comets out there and they don't know where most of them are. And the big ones don't have to hit Earth for humanity to be staggered. An asteroid as small as 40 yards can level a city and incinerate millions of its inhabitants in the process. These asteroid trackers are the watch guards against killers from the cosmos and when they say they are underfunded and understaffed we might want to sit up pay attention.

In 2016 NASA announced the creation of the Planetary Defense Coordination Office. One of its purposes is to track NEOs larger than 60 yards and come up with a response to the threat and a plan to mitigate the damage. In October of 2002 the B612 Foundation was formed. It's a private group headquartered in the United States with the single focus of defending the planet against asteroid strikes. It is comprised of former NASA astronauts, scientists, engineers and other like-minded individuals. It is an impressive collection of brain power and experience whose predictions are more than worthy of consideration. One such prediction is particularly concerning. It has revised the time period in which asteroids that are called "city killers" impact Earth. The original estimation was these "city killers" impact Earth every one hundred years or so. The new estimation might be more than five times more frequent than what was previously thought. One of the foundations plans is to build a space observatory that would be privately financed. The observatory would be called the Sentinel Space Telescope and is scheduled for launch in 2019. The estimated cost is 450 million dollars. The telescope would last for about a decade. Fund raising has been lower than expected which highlights the need to increase public enthusiasm and participation. Scientists hope the Sentinel Space Telescope is the solution to problems caused by the blind side. Telescopes cannot look into the blind side because it is in the direction of the sun and we can't look into the sun. The Sentinel will be launched to a certain position in space where the sun will be behind the telescope instead of in front of it. The sun is currently in front of every telescope on Earth that searches for asteroids which is why there is a blind spot. The Sentinel's position in space should eliminate the blind spot. Small asteroids are also hard to see because they don't reflect much light but the Sentential has an infrared camera that can see in the dark by identifying the heat instead of the light that an asteroid gives emits. Sentential is scheduled for launch in 2019, so in theory there might be just a couple of years left where we have to continue to beat the odds posed from asteroids that come from the blind side.

NASA recently terminated its relationship with the B612 Foundation and created something called the Planetary Defense Coordination Office. Its mission is much like that of the B612 Foundation except the Planetary Defense Coordination Office will carry on the work through government agencies instead of private sector groups. Members in the group report that Earth is not in immediate danger from an asteroid impact but that is of little consolation when considered there are numerous asteroids that we don't know about, including the ones that come from our blind side and that's not to mention the ones that are too dim to see until they're nearly upon us.

An examination of the recent record will confirm how vulnerable we are to an asteroid strike, remember the Chelyabinsk incident in 2013 or the close call from the Halloween Asteroid in 2015. The Halloween asteroid flew by Earth at the relatively safe distance of 300,000 miles. The Moon is about 230,000 miles from Earth, so it was not a close as many of the others. The Halloween Asteroid is about 1,300 feet wide and would surely cause death and destruction if it were to impact Earth. The Chelyabinsk asteroid really came out of now where and in no way was predicted. It blew up in the sky almost 15 miles above Russia in 2013. Its explosion was 30 times brighter than the Sun and had it impacted Earth it would have killed millions of people with a release of energy said to be 30 times as powerful as the Hiroshima bomb.

June 30th is Asteroid Day, a day set aside to talk about the inevitable. A panel of experts is scheduled to attend and discuss ways to better identify asteroids and stop the ones headed our way. One of the event organizers is a former rock and roll star, the one-time lead guitarists from the band Queen. They will talk about some of the obvious subjects: sooner is better than later when it comes to asteroid detection, reason being the farther out the asteroid is, the is easier it is to move. They will also probably talk about the "Hollywood Approach" to the problem, depicted in a big screen block buster in which humans sent a nuclear bomb to destroy the global extinction asteroid bearing down on Mother Earth. The approach made for a good movie but we apparently can't do that just yet. Scientists say even if the technology were present the "Hollywood Approach" would only be a last resort when there is little or no warning, and again it's still not possible.

Two of Many!

One of humanity's greatest mechanical achievements has to be the Voyager I Space Probe, mentioned earlier it was launched in the fall of 1977 and is still on the job, last thought to be just outside the solar system, billions of miles from home. Voyager I and its successor Voyager II have traveled to all of the giant planets in the outer solar system, from Jupiter to Neptune. To get up to speed for the long journey Voyager I used the gravity assist procedure where it would fly by a planet and then use the planet's gravity to accelerate in a sling shot fashion. It has been hurtling through space at about 39,000 miles an hour for the last 35 years. In that time is has traveled 11 billion miles, enough to circle Earth more than 400,000 times. It was originally designed to survey only Jupiter but has since surveyed all of the distant planets in the solar system as well as their Moons. It is powered by plutonium and shuts off everything but the essentials during travel to save energy. It communicates through radio waves, regardless of the distance. Radio waves can travel at the speed of light, enabling Voyager I to send back data to Earth in about 16 or 17 hours from that distant location.

The probe also carries a message for alien life but if any alien life has been observing human behavior then that message would be a misrepresentation. The message has greetings in 55 languages along with pictures of different life forms on Earth as well as songs and other material thought to represent humans. It is called the Golden Record. The contents were selected by a committee chaired by Dr. Carl Sagan and include a message from the US President at the time. Dr. Sagan was a scientist best known for getting the American public interested in astronomy by explaining complicated matters in a simple and entertaining fashion.

The Golden Record contains 115 images and pictures most of which depict the pleasant side of human life. It basically contains multiple greetings in several languages. The first language presented was from the Sumerians some 6 thousand years ago and ends with a dialect of modern Chinese. The Golden Record carries no references to the true nature of our species, nothing about our violent, racist, sexist nature that have produced numerous wars which have marred our existence since recorded history. There is no reference to how material things such as money, cars and clothes preoccupy much of our time at the expense of pursing compassion and understanding. Not a one of our shortcomings was mentioned in that representation which raises the question is the Golden Record an honest depiction of the human character?

There is conjuncture that long after humans have gone extinct or have left the planet, the former is probably more likely than the latter, Voyager I will continue to stay on the job, who knows how far out, its batteries long since dead, losing its ability to send data back home, left on its own to explore the unknown with no home base in existence to receive its valuable information. In fact, one of the Star Trek movies was based on Voyager. The Enterprise was threatened to be destroyed by a superior alien craft called VGER, until crew members from the Enterprise figured out that VGER was actually the probe Voyager (actually Voyager 6 in the movie) which had been traveling the galaxy for years and needed someone to figure out its code which allowed VGER to release all of the intimation it had gathered. Only VGER's creator had the password to the code. The crew of the Enterprise retrieved the password from its archives, VGER realized it had found its creator and did not destroy the Enterprise and released its wealth of information. VGER was short for Voyager but years of wear and tear in space obscured the missing letters that completed the word Voyager and instead only left VGER.

Another one of humanity's greatest mechanical achievements has to be the New Horizons Space Probe. It recently flew by Pluto some 4.5 billion miles away and has the distinguished honor of being the fastest space probe ever launched at 36,000 miles another. Voyager is about three thousand miles an hour faster but Voyager reached its top end with the help of a gravity assist from Jupiter. That gravity assist boosted Voyager's speed by nine thousand miles an hour and without that it would have taken another three years to reach its destination. New Horizons was so fast that it made to the moon in about nine hours. It took the Apollo missions and about three days to make the trip, a distance of 238,000 miles. When the probe was launched Pluto was still a planet and while in route Pluto was demoted to a dwarf planet during the nine-year trip. Either as a sign of protest or appreciation, New Horizons carried the cremated remains of Clyde Tombaugh. Mr. Tombaugh was the astronomer who discovered Pluto in 1930. He died in 1997 and a portion of his ashes were placed inside New Horizons for the trip to Pluto. Out of respect for the deceased, there will be no speculation as to what Mr. Tombaugh might have thought when his remains landed on Pluto which was a planet when the journey began but was demoted by the time he arrived.

New Horizons is power by nuclear fuel which powers a contraption called a radioisotope thermoelectric generator. Solar panels would probably have been a more efficient source of energy but the problem was New Horizons was too far from the sun for them to work. Its maximum output is about 300 watts which is about the same amount of energy used to power the average blender. So New Horizons is about 4.5 billion miles away blasting through space at some 39,000 miles an hour with a motor like a kitchen blender's.

New Horizons might be able to stay on the job if it's in the budget. While New Horizons is now past Pluto there is the possibility that it can still work and collect data if enough funds can be raised to explore the Kuiper Belt. The Kuiper Belt is thought to be up to four billion miles away and is full of asteroids. Many of those asteroids are thought to be remnants from the Big Bang and contain information that could tell us more about our origins and the early universe, so if we can afford it we might learn more.

Two in One Year!

The space exploration community lost two of its family members in one year. 2012 saw the passing of Sally Ride and Neil Armstrong. Ms. Ride was the first American woman in space and Mr. Armstrong was the first person to walk on the moon. Ms. Ride made history in 1983 and Mr. Armstrong made it in 1969.

Neil Armstrong is probably the most famous astronaut, period. His superior notoriety was probably a direct result of being the first person to walk on the moon but there is more to his story than just the headline. He was born in Ohio and was interested in flight almost from the beginning. When he was two his father took him to a local air plane race. Three years later his father took him on his first plane ride. Neil knew how to fly a plane before he could drive a car. Obviously he was an extremely gifted individual who made the most of his opportunities and natural talent. He excelled in academics, receiving a scholarship in aeronautical engineering, attending Massachusetts Institute of Technology. He later flew some 78 missions in the Korean War and received the Air Medal for 20 combat missions. He would continue to excel in the military as a pilot and engineer and his accomplishments are too long to list here. We would be remiss not to mention at least two of the many occasions when Mr. Armstrong cheated death.

He was just 21 years old and was flying right into combat in the Korean War. After he was launched from an aircraft carrier he took out a strategic bridge in North Korea. On the way back to the ship his jet struck a cable in mid-flight. The collision could have easily killed him. The cable cut off several feet of the plane's wing. The young Mr. Armstrong stayed cool and fought with the plane which did not respond. He continued to fight and wrestle with the flying death trap until he was able to eject at high speed and land in friendly territory. He surely had stories to tell that night but they would pale in comparison to his tales that would come.

In 1962 government officials asked Mr. Armstrong the obvious, would he like to join NASA and become an astronaut. The answer was an immediate yes and he moved to Houston, TX to join the program. Four years later he found himself as the Command Pilot for the Gemini 8 mission. It was on that mission that all of his combat experience, academic training and natural aptitude for flight became apparent. After he successfully completed the first ever docking mission between Gemini 8 and an engine booster the two connected space crafts tumbled into an uncontrolled roll. Gemini 8 was un-docked as part of an emergency procedure but it only spun faster. At one point the sizable space craft was spinning so fast that it was revolving once every second. The G forces were too great for humans to last long and Neil was threatened with blacking out. Without much time to come up with a solution, Neil Armstrong managed to keep is cool, along with his consciousness and perform a maneuver that saved the day. He used the re-entry thrusters which were blasts from the engine to stop the craft from spinning, he aborted the mission and proceeded with an emergency splash down. The chaotic event was classified as the worst uncontrolled spin on record. Survival of the spin was one more time when Neil Armstrong cheated death.

As a youngster Neil Armstrong had many odd jobs including walking among the dead. When he was ten he mowed the lawn of the cemetery in his home town. Apparently he was paid by the job and not by the hour and received one dollar to mow the entire grave yard. He used the job to pay for flight lessons. Seems he might have had to mow the cemetery nine times just to pay for a one-hour flight lesson. He later got a job at a local airport where he helped out with just about any and everything, his energy and enthusiasm were so great that they earned him rides and flying lessons. From the earliest he earned the reputation of being able to fly anything, including the dangerous rocket plane the X-15 which killed a pilot who was training to become an astronaut.

In the summer of 1969 Mr. Armstrong and crew took center stage while the world watched, an estimated 600 million observers on Earth looking on or listening in amazement as the craft blasted toward the Moon. They would travel 238,000 miles in 76 hours to reach the lunar surface. Armstrong told mission control "The Eagle has landed" and a short while later came those words that seem to last forever, "That's one small step for man, one giant leap for mankind." All this was done under what has been called the largest audience ever assembled on Earth for a single event.

When he came back to Earth he was a hero, an instant celebrity. When everything settled down, Mr. Armstrong bought a farm in Ohio, became a professor and despite the circumstances and notoriety did an amazing job of leading a normal life. After retirement he was the target of a hoax that claimed Neil Armstrong had converted to Islam while on the moon. Mr. Armstrong kept his religious beliefs a secret but when he died he wanted to be cremated which is forbidden by Islamic law. He passed away from complications of cardiac surgery to relieve blocked coronary arteries in the summer of 2012.

Sally Ride was an American physicists and astronaut who was born in Los Angeles, CA in 1951. She joined NASA in 1978 and became the first American woman in space in 1983. At one point she was the youngest astronaut to enter space at the age of 32. As you can imagine she was a gifted student but was also a nationally ranked tennis player. When she was a student at Stanford University she was one of 8,000 people who responded to an ad for applicants for the space program. Her first space flight interview might have been the most taxing from a mental perspective as she had to tolerate sexist's questions like, will the trip in space effect your reproductive organs, and when things go bad do you cry. She was a professional, kept her poise and responded with grace, patience and tolerance.

Sally Ride took part in several space flights and logged more than 300 hours in orbit. One of her jobs on board the shuttle was to work the control arm that put satellites in space. She stopped working for NASA in 1987 and in 2003 she was inducted into the Astronaut Hall of Fame. After retiring from NASA Sally Ride pursued her personal interests including work as a professor, an advocate for international security and arms control and teaching children about the wonders of science in general and space exploration in particular. Sally Ride's advocates say she did not sell out by trying to turn an easy dollar after her NASA career and instead stuck true to her core values. One publication wrote that after her two space flights she was one of the most influential woman on Earth and had ample opportunity to exploit her fame through the many endorsements and movie deals that came her way.

She focused her attention on endeavors that contained substance such as educating children in general and encouraging girls into the sciences in particular. When she was 18 years old she was ranked the 18th best player in the nation with so much promise that tennis great Billie Jean King encouraged her to turn pro. She died in 2012 from pancreatic cancer at the age of 61. No doubt her impact will be felt for generations to come in the children and girls who were encouraged to follow in her footsteps.

It's Like That!

So what is life in space really like? Only about 536 people can answer that question including, Scott Kelly, Marsha Ivins and Doug Wheelock. Mr. Kelly has spent more time in space than any other astronaut, more than a year and apparently he's ready for more. He said if he had to, he could go another year. Those comments were given during an interview from the International Space Station just a few days before he is scheduled to come home.

Does Scott Kelly miss home and other people? The answer to both questions is yes but he sees his job as a tremendous privilege and the time away seems to be a small price to pay for that valuable opportunity. At the time of the interview he had just turned 52 years old and he looked forward to getting back to Texas and jump in his swimming pool. He has two daughters and a girlfriend waiting for him. He also has a twin brother who is a retired astronaut. Scott has been to the International Space Station twice and traveled in orbit several other times so when he returns home he will have spent more than a year in space, 520 days to be exact.

He admits that space is a harsh environment where he never felt normal. There is the effect on personal hygiene because there is no running water, which was compared to camping in the woods for months on end. Then you might have to deal with claustrophobia, living in cramped quarters for extended periods of time with several other astronauts and cosmonauts. His working quarters might be even more cramped, which have been compared to working marathon shifts inside a phone booth and his sleeping quarters were no bigger.

American Astronaut Scott Kelly returned to Earth in early March of 2016 after spending 340 consecutive days in space. He had to be helped out of the craft and looked to be in visible pain, no doubt suffering the debilitating effects of nearly a year in a zero gravity environment, including loss of muscle mass and bone density to name a few. He unceremoniously touched down in Kazakhstan because remember the Americans have no way of returning its astronauts back to Earth, or for that matter sending them into space. That's one of the consequences of decommissioning the American Space Shuttle Program. Another consequence is having to rent seats on a Russian space craft in order to get to the International Space Station: at last check they were 82 million dollars a seat, up from 71 million.

Mr. Kelly was also two inches taller when he came home which is not uncommon. Seems when humans are in space for extended periods gravity is no longer compressing their spines and the spines become elongated which makes the humans taller. So for a while he was taller than his twin brother. It took Scott about 28 hours to return to his normal size which makes you wonder could he actually feel his body shirking as he lost about a quarter inch in height every three hours or so.

Doctors will also medically compare the twins, one in space and the other on Earth to determine if there are any significant differences. Of course hygiene was one difference, Scott jokingly said he hadn't showered in a year, which might have been true. His mental health will also be examined as well as any possible effects that long term exposure to radiation in space might have had on his body.

There's also the isolation of space, sure he might have three of four other people crammed inside the station with him, but there's no denying the fact that he's isolated. Of course there is that amazing view of the Earth, said to impact astronauts in unexpectedly profound ways, but the benefit from the view can only go so far. And then there's answering the call for national duty in the interest of advancing science and the satisfaction of a job well done. Only four other people have spent more time in space and they're all Russians.

One astronaut said one of the tensest periods in space was right before he was to return home. After spending a total of 178 days in space spread out over two missions, Astronaut Doug Wheelock said the anticipation of the return trip home was an intensely bitter sweet moment. One the one hand he just counted down the hours throughout the night, thinking about what he would do and who he would see once back on Earth. But on the other hand there was an element of melancholy, tempered with sadness and reflection as he was about to leave his temporary home in space that protected him from the lethal vacuum just outside its walls. But like most people in space when their time comes they are more than ready to go.

Astronaut Marsha Ivins, in the online publication wired.com, said the realization of looking down on Earth from space and realizing that you are not on the planet is literally an out of this world feeling. She too described the accommodations as cramped, crowed and at times uncomfortable but that view made up for some of the discomfort. While she said space travel was not glamorous she did say it was breath taking. She apparently never ate much on her several flights as nausea was a common ailment for astronauts. While many astronauts and cosmonauts never fully adapt to a zero gravity environment most learn to cope with in a few days.

She flew five missions which included duty on the Space Shuttle Atlantis and like many others said sleeping was a problem. She said you climb into a sleeping bag stuck to the side of a wall or ceiling and try to rest. She would then tighten straps to make her feel tucked in and then would try to position a pillow in the right place under her head and try to relax the neck. If the arms struck out they could disrupt sleep because those appendages would float around. All of the astronauts might be compared to explorers or mountain climbers or travelers on the ocean, who have many hours alone with their thoughts all the while trying to stay focus on the job at hand. Despite all of the problems mentioned she made sure to point out that all five of her space flights were relaxing and that's despite the long hours, the heavy work load and the intense pressure of not making any mistakes. She said she would go back in an instant.

The trip home can be more radical than the one into space which is something else that also occupies the thought of astronauts waiting to return to Earth. The return trip has been described as lesson is sensory over load, ripping through the atmosphere at some 16,000 miles an hour, a ride so extreme that rough and bumpy are inadequate adjectives. There's also the intense heat, coupled with the thought that as hot as it is inside, the heat outside would kill you.

And there's more, that tremendous gravity, its relentless and suffocating pressure, its g-forces always pushing down. There also what's called the moment of truth when the craft lands. If the crew was in a shuttle that moment of truth might come when the craft breaks through several layers of atmosphere, and has not broken apart or lost any heat resistant tiles and crew members can clearly see the surface and maybe even the landing strip far off in the distance. For crew members that return on the Russian Soyuz Space Craft the experience is said to be a little different. One crew member reported seeing the outside of the heat shield actually melting away from the space craft. The landing has been described as a "rough trip", apparently one craft hit a couple of times and then rolled over, after that there can be about a 20-minute wait before the latch is opened and the second moment of truth comes.

The second moment of truth is when astronauts and cosmonauts try out their legs. Almost everyone returning from space reports some sort of shakiness once their feet are back under them on Earth. Depending on the length of their stays in space the astronauts and cosmonauts have undergone varying degrees of bone loss and muscle atrophy. It's as if the brain has told the body that the legs were not needed in space and now back on Earth those idle legs must be deployed back into action, but you don't know how they will work or how long it will take them to return to normal.

Once the legs have started to return astronauts and cosmonauts alike report a renewed awareness of their senses. In space senses like taste, smell and balance are dulled. One astronaut was disappointed after taking some delicious high priced chocolate into space only to find it it tasted like wax. The saying there are no good meals in space, only good views seems to ring true which could explain why people returning from space often chart a direct path to the nearest bacon cheeseburger. Others talks about the smell of freshly cut grass or the smell and sound of surf lapping on the beach. Sights, smells and sounds that were once taken for granted become intoxicating once astronauts and cosmonauts return home.

No doubt astronauts are worried getting trapped in space with the lack of pressure having fatal results

Take water for example, it expands into vapor in space because there is not enough pressure to keep it liquid. The body is about 70 percent water and would have an adverse reaction if humans were exposed to the vacuum of space. No pressure in space can also means no pressure inside the human body which is a condition called Hypoxia. During Hypoxia, oxygen in the blood escapes which leaves the muscles, joints and vital organs with no way to function and they can just stop working. A heart attack might follow because pressure has dropped so low blood is no longer pumped to the heart. Boiling points are also effected by the surrounding pressure. The lower the pressure means the lower the boiling point. So in space the boiling point of any liquid could drop as low as the temperature of the human body which would make the blood boil. But because the skin is remarkably tough the blood would have to be exposed to the vacuum of space for it to boil.

Exposure to the sun light in space would be extremely painful, with no atmosphere or ozone layer to filter out harmful ultra violet rays any exposed skin would be badly burned and could lead to skin cancer. If you looked into the sun under these conditions, you might become blind when its bright rays hit the sensitive retina of the eye. Gamma rays, protons and other sub atomic particles can easily damage the DNA of humans with possible fatal consequences. Despite all this and more the majority of astronauts surveyed said they would return to space in an instant.

Time for a Shower!

The Perseid Meteor shower has been called fine family entertainment free to all who care to observe. The Perseid is a collection of debris, little bits and pieces of rock and metal left behind from a comet called Swift Tuttle. Earth passes through the comet's debris each time our planet orbits the Sun which is why the meteor showers occur once a year. Swift Tuttle was first discovered in 1862 by American Astronomers Lewis Swift and Horace Tuttle who might have been able to name it after themselves. Mr. Swift first spotted the comet at the height of the Civil War. A few days later it was spotted by Horace Tuttle. Years later it was learned that Comet Swift Tuttle comes by every 134 years or so depending on who calculates the estimations.

All of this free family entertainment is displayed every August and in 2016 is projected to peak in the middle of that month. If the conditions are just right with clear, dark skies up to 100 of these meteors can been seen with the naked eye every hour. Again these meteors are very small, most are the size of grains of sand, some are bigger, like the size of a pea or marble, most burn up in the atmosphere and never make it all the way down to the surface.

We're told Swift Tuttle won't hit us even though it does fly by from time to time. Its next visit is scheduled for 2126. The last time it came by was in 1992. That fly by was an uneventful visit as it was too far away to be seen with the naked eye. Prior to that was in 1862 and if the President at the time gazed at stars then Abraham Lincoln was treated to fine family entertainment free to all who care to observe.

Scientists tell us when Earth rotates it reaches a certain position where part of the planet attracts more debris than the other part, which is why dawn is the best recommended viewing time. The little meteors are left behind like a trail as they drift from the tail of the comet. They are not packed tightly together but instead are separated by 50 or 100 miles. Each time the comet passed it left behind earlier trails of debris which remained in their same orbits for years, just waiting to be illuminated for our viewing pleasure. Or put another way, the Comet's tail just sits there in space and once a year Earth passes through the tail as part of its annual journey around the Sun. When the particles hit Earth's atmosphere they light up.

The particles that make it into Earth's atmosphere are moving fast and glow brightly because of the intense atmospheric friction, which in part explains why objects that are so small and so far away can be seen with the naked eye. They can be as hot as 3,000 degrees, made possible by their tremendous speed, estimated to be in excess of 133,000 miles per hour. The high speed of those particles might be explained by considering the speed of Earth. Our planet travels around the Sun at about 67,000 miles an hour. We're told the comet's residual particles enter Earth's atmosphere at 133,000 miles an hour. Could the particles actually be traveling at 66 thousand miles an hour and when they hit Earth which is moving at 67,000 miles an hour, the combined force of the impact creates the equivalent of an impact at 133,000 miles an hour, similar to two cars traveling at 30 miles an hour each and collide in a head on collision which boosts the relative speed to 60 miles an hour?

Most of these little meteors become visible around the magical mile maker of 62 miles up, the agreed upon altitude where space begins. Some of the bigger ones make flashes called fire balls and can actually be heard exploding from the ground. Imagine hearing something 60 miles away. Most of the particles never hit the ground but when they do they are called meteorites and when they're racing through Earth's atmosphere they're called meteors and when they're in space they are called meteoroids. As previously mentioned I don't know why scientists felt compelled to come up with three names for one rock.

The Comet Swift Tuttle is the largest comet known to make repeated passes by Earth and we are fortunate that it will not stop for a visit because it is about 16 miles across, more than twice the size of the one that did in the dinosaurs some 65 million years ago. The Chinese were apparently the first to see the comet back in 69 BC, which helps to explain why astronomers know so much about it. If it were to hit Earth it would likely cause destruction on a global scale and might even cause the planet to go extinct. The energy released could be 27 times more powerful than the Cretaceous Paleogene impactor. That was the asteroid believe to have struck 66 million years ago and killed most of the life on Earth, wiping out about three quarters of plant and animal species including the dinosaurs.

One of the many things that make comets so important is they're probably the only objects old enough that were made from the same material from which the universe was formed. Our Earth, its Moon and other bodies have all been changed over time in some shape or form, either by erosion, multiple collisions or the movement of tectonic plates.

There are two basic types of comets, long period and short period. It takes on average about 200 years for long period comets to orbit the sun and short period ones make the trip is less than 200 years. Most comets are thought to come from the Oort Cloud which might contain about 100 billion of them. Many of the large ones, like Haley leave behind a trails of debris in their orbits which creates annual meteor showers like Perseid. There was considerable panic in 1910 when Earth was ready to pass through the trail left behind by Haley's Comet because people thought they would be in harm's way. Nothing happened other than they were treated to fine family entertainment free to all who care to observe.

Beauty and danger often go hand in hand as Comet Swift Tuttle has be described as the single most dangerous object known to humanity. Who knew that a producer of fine family entertainment free to all who care to observe could be so threatening. Consider ourselves fortunate that its projected close encounter with Earth won't be until the year 4479. So what are the chances of a direct impact on Earth or even the Moon? They are calculated at one in one million, long odds made more perilous by what's a risk.

62-Mile-High Club!

If you are not familiar with the mile-high club it seemed to be a risky practice on board jets before the 911 crisis tightened airport security across the industry. Entry into the club only required two people have sex on a plane. Considering how sex is an integral part of the human existence and how it is central to our core behavior, it might be only natural to ask if humans have created a 62-mile-high club. 62 being that magical mile marker where space actually begins.

For the purposes of a hypothetical scenario let us say sex could include any number of frisky behaviors including intercourse. So sex on board a shuttle would be hard. For one it is cramp with little or no privacy. NASA says it never happened and astronauts past and present don't talk about it. The space shuttle is about the size of a Jumbo Jet but the shuttle has nowhere near the open space of the commercial airliner. The shuttle has two main compartments which are about the sizes of small bedrooms. The bathroom is even smaller, made up of nothing more than a toilet and a curtain. That's about it, there are no closed rooms where a couple could hide and have sex. Of course that's not to say that sex on the shuttle never happened, only that if it did happen it would not have been easy and NASA won't acknowledge it and astronauts don't talk about it.

The International Space Station provides a little more room and a little more opportunity. If there is crew of three people, they will often split up when they go to sleep. Two people might head off to two small sleeping cabins at one end of the station and the third person might crawl into a sleeping bag that could be more than 100 feet away. The beds inside the sleeping cabin come equipped with straps and special designs that would make sex difficult.

Then you have to consider the element of energy, which has been a problem for couples on Earth let alone in space. Astronauts have demanding schedules and long hours. The heavy and stressful workloads can leave them with little time nor energy for getting frisky. But they do have weekends off when they can watch movies, read books and in general just enjoy themselves and in this context the term enjoy is used in the broadest of terms. And after reviewing more documentaries of the ISS and the additions made to the station it becomes apparent that its many compartments, extensions and connecting tunnels might provide a place, even in for a short period of time in those cramped confines where two likeminded individuals could engage in behavior that could be considered frisky.

Then there are physical logistics of having sex in space where there is no gravity, things like food, tears and other liquids just float around in the cramped confines of the closed cabin. There have been stories were astronauts have had accidents that resulted in some of their excreted wastes just floating around the cabin, offensive but true. There are also cases of how food that was not issued by NASA was smuggled on board and once it was eaten the crumbs floated around the cabin and threatened the clog some of the instrumentation. Remember how difficult it is to go the rest room, each astronaut is provided with a special device that is custom fitted to her or his genitalia into order to prevent their waste from escaping and floating around the cabin. But as one anthropologist said when it comes to the human behavior, where there is a will there is a way and one should not underestimate human ingenuity.

Then there are the pragmatic elements of having sex in space. How does the lack of gravity effect an erection? Well apparently not very well. According to the experts, in an environment with gravity like

Earth blood flows to the lower parts of the body but in space blood rises to the head and chest and not to the penis which would make getting an erection difficult but not impossible. Also in a weightless environment every push or thrust would immediately propel one's partner in the opposite direction. One of the requirements for sex is the coupling behavior which might never be accomplished in space again because each movement of passion would cause the partners to move away instead of together. One possible remedy might be some sort of anchoring device which in all likelihood would be awkward and might cause the couple to lose the mood.

Women would have the same problems even though they have different plumbing. When they get frisky the blood rushes to their genitals which causes the required lubrication and swelling which might not happen in zero gravity. Its common knowledge that testosterone increases the libido or sex drive but for some unexplained reason testosterone levels are reduced in space but return to normal when astronauts are back on Earth.

Once instance when sex in space might have occurred was when NASA sent a married couple into orbit. The administration usually prohibited that sort of thing but it made an exception in this case. The year was 1991 and in one picture the husband had his arm around his wife but as you can imagine both people refused to answer any questions about their relationship while they were on board the flight. But was it more probable than not? They were basically newlyweds, having wed very close to the scheduled launch. They were presented with an extraordinary view, think of what motivation views from a mountain top or lake side provide for couples, married or not, and now you have these amazing views from space, enjoyed by young newlyweds, again what is more probable than not?

The year before a pair of Russian Cosmonauts on board the MIR Space Station stoked the rumor mills. They were recorded on video where one was splashing water on the other in what appeared to be a playful motion, described by one writer as an unorthodox coziness. When one considers all of the strange and bizarre places on the planet where couples have had sex, one might assume that our natural behaviors on Earth will continue in space.

Can't Make This Stuff Up?

In 1969 the American space program was still in its infancy and its president wasn't sure if their very big and important space flight would succeed or fail, either way it would do so in front of an international audience. President Richard Nixon gave the astronauts of Apollo 11 a personal send off and wished them all the very best. But just in case the astronauts encountered the very worst and were killed, the president had a farewell speech prepared. He didn't write it but here's an excerpt: "Fate has ordained that the men who went to the moon to explore in peace will stay on the moon to rest in peace. These brave men, Neil Armstrong and Edwin Aldrin know that there is no hope for their recovery. But they also know that there is hope for mankind in their sacrifice. These two men are laying down their lives in mankind's most noble goal: the search for truth and understanding."

Seems lots of folks had a bad feeling about the Apollo 11 mission, not only the president who felt compelled to commission a backup speech in case of the crew's death but also insurance companies. There are reports that NASA searched far and wide for a policy on the crew but no company could be found that would write such an out of this world risk.

There was the unspoken thought that this was going to be a suicide mission and the government had no way in which to compensate these brave young souls who were prepared to lay down their lives for their country. Something had to be done. Turns out NASA had the astronauts sign a series of their photographs which it would sell in case of their deaths. NASA hoped that the proceeds would provide adequate compensation to their families as well as leave behind some sort of legacy. Fortunately, it was never needed. We might never know how many autographs were signed or how much money could have been raised but years later a few of those autographed pictures were sold for tens of thousands of dollars at auctions.

Apollo 13 command pilot Jack Swigert maybe best known for uttering those famous words "Houston we have a problem." Well Mr. Swigert had another problem while in flight, he forgot to file his taxes. Some of the folks at mission control and others on the ground thought it was funny and made light of it with comments like Swigert's tax return will be used to buy fuel, possibly as some sort of penalty for filing late. Commander Swigert didn't think it was funny, maybe realizing the nature of the I.R.S. and didn't drop the matter until he was given an extension. Apparently he was well acquainted with the two things that are certain in life. A little more than three decades later a Russian Cosmonaut had the same problem but was able to pay in space, through the internet and became the first person to do so.

NASA might want to improve its employee background checks after the antics of one of its astronauts. After embarrassing herself and the administration, former Astronaut Lisa Nowak was charged with the attempted kidnap and murder of the girlfriend of an astronaut with whom she was having an affair. Long story short, Ms. Nowak drove have half way across the country with a NASA diaper on so she did not have to stop the car to go the restroom when nature called and instead could relive herself right there behind the wheel in the front seat while still on the road.

She made an all-night b-line to the competition's house where she was armed with a bail peen hammer, and a BB gun and other items ready to snuff out her romantic rival. When she was arrested by police Capt. Nowak was wearing a wig and a trench coat, she had taken off the diaper. After just driving 950 miles you can probably guess if the diaper was used or not. The incident happened in the winter of

2007. She was fired from NASA in the spring of that year and two years later plead guilty to the lesser charges of felony car burglary and misdemeanor battery. One can only assume she underwent some sort of mental health counseling before fading into obscurity.

Apparently there was car of choice for the astronauts of the 1960s and it wasn't a family sedan. It seemed to be the Chevrolet Corvette, the auto giant apparently saw the astronauts as the perfect marketing tool for its top end car. NASA might have paid its astronauts back then about 17,000 dollars a year which works out to more than 100,000 dollars in today's economy. While the astronauts of the early 1960s were not getting rich they did enjoy some of the perks of the job. One astronaut showed up to training driving a late model 1950s corvette and others soon followed suit. General Motors offered friendly lease terms that some observers say were almost free. NASA did not endorse corvettes but its astronauts certainly seemed to. At one point so many astronauts drove corvettes that the media stated tthe astronauts helped to build the corvette's reputation as the national sports car. Many astronauts had more than one corvette and one astronaut had ten. Out of fear that NASA might somehow appear to be aligned with General Motors, the space administration brought an end to to the relationship and all of those great deals by the early 1970s.

The Future's Already Here!

It is hard to believe but part of the future is already here, it might have arrived in 1998 when the International Space Station was launched. Regardless of humanity's next endeavors in space, chances are the ISS will play an integral role. It can be used as a launch pad for significant explorations into space. It is already being used as a laboratory to conduct tests in the weightlessness of space. More than 65 nations have sent their scientists to conduct more than 1,500 experiments in that station's zero gravity environment, experiments that could not be re-produced on Earth.

Technology gained from the International Space Station can help with water purification. Astronauts are taught the latest methods to conserve, filter and purify water on the station and that knowledge directly benefits almost everyone, especially people in third world countries and elsewhere who dealing with drought and other water shortage problems. The process is so refined that astronauts on the ISS drink their own urine. Of course that's after it has been purified and the procedure highlights one of the differences between the Americans and the Russians. The Russians won't drink their urine and who could blame them. The Russians have a separate water purification system and who could blame them. There was no explanation for the Russian refusal and who can blame them. The taste of the recycled, purified urine is said to be comparable to bottled water with the biggest obstacle to overcome being the psychological factor of what's being ingested. The Americans also drink something called condensate which comes out of the air and might be even more offensive. Condensate is a concoction made from sweat, breath residue, shower run off and urine from animals on board. The process is so refined that more than 90 percent of the water on the International Space Station is reclaimed with some of the Americans occasionally drinking urine from the Russians. One can imagine the astronaut's behavior provides the Russians with ample material for comedic relief at the expense of the Americans.

Work on the Intentional Space Station can also help people see better on Earth. The human body contains more than 100,000 proteins and billions more are found in space which might be the perfect environment for proteins to grow. Evidently micro gravity allows for optimum growth for proteins which can lead to the development of new medical treatments. Proteins are one of the basic building blocks of life. Work on the station has also developed stronger vaccines and research indicates that certain bacteria such as Salmonella might become more virulent in space. Salmonella sickens and kills hundreds of Americans annually. Researchers also study possible causes of why Salmonella might become more active in space which could lead to new microbial vaccine development. Results from using the robotic arm on the ISS has been applied to clinical trials in the treatment of breast cancer cases on Earth. The machine from the ISS can help identify the size and location of a tumor and assist in precise movements during biopsies. Technology developed on the station is beneficial in laser eye surgery in that precise movements used in space allow surgeons on Earth to track the position of the eye with unparalleled precision. ISS studies on bone loss also help doctors to better understand the mechanics of osteoporosis. Osteoporosis causes bones to become weak and brittle. The disease is extremely painful and periodically causes death when it's associated with hip fractures in the elderly.

The You Tube Space Lab Competition in conjunction with other student programs inspires and motivates millions of students across the globe by allowing them to compare their experiments in the classroom with similar ones conducted on the International Space

Station. Results from both experiments allow students to improve their skills in math, science, engineering and technology which prepares them for careers in those disciplines. It is also used as an important galactic observatory and it provides valuable data regarding the prolonged effects of human exposure to radiation without the protection of the ozone payer or our atmosphere. It is truly international, 15 nations took part in its construction which can house up to six people and is the world's first and only permanent occupied facility that orbits Earth.

Some astrophysicists say if we have to crawl before we can walk, then the International Space Station represents that walking stage. The huge facility allows us to learn more about Earth's low orbit which is paramount to discovering the secrets of what is beyond in the great abyss. The station has been called the Gateway to Space. Earth's low orbit begins at the surface and extends about 1,243 miles up. Objects that are below this altitude undergo quick orbital decay and begin to fall back to Earth. A tremendous velocity is needed to maintain this orbit, said to be about 17,672 miles an hour. The higher a vehicle goes and the less speed it needs to maintain its altitude and orbit. So we can learn much about the dynamics of deep space without actually traveling there. Low orbit is also the cheapest way and place to deploy satellites which is another reason why the area is so important.

The International Space Station might also help us get to Mars by helping to speed up communication between astronauts traveling to or from Mars and Earth, which could be of significance in the case of a medical or mechanical emergency when time is of the essence and could actually save lives. The station is also being used to study and understand the psychological impact of confinement and isolation in space. Researchers have studied the journals of astronauts and former astronauts who have been to the ISS to help them with this work. There are more than a 1,000 books left behind by these astronauts with a vast range of detailed accounts regarding emotional and psychological states and behavioral observations during extended stays on board the ISS.

The future of the ISS seems to be bright but short. The United States recently announced it would only commit to the project until 2024. That deadline might come sooner than later depending on the what the other contributing nations do. Remember the ISS was an international collaborative effort and the withdrawal of other nations could cripple or derail the project. Some national representatives have already asked; why should tens of millions of dollars continue to be invested in the ISS if it's only going to be around for another 10 years or so? The counter argument is, keep the ISS working until at least the year 2050. Advocates say that would give scientists the requisite time to complete some of the important work that has already begun. One example would be the research of learning more about dark matter that began on the ISS. As mentioned earlier, dark matter is thought it make up almost 30 percent of the entire universe with scientists saying research conducted on the station cannot be duplicated on Earth.

They are also conducting vision tests on astronauts, sight is all important in space and is impacted by the effects of zero gravity. Those effects might be amplified during long term flights and could also affect vascular, central nervous and immune systems. Other tests on the ISS are designed to determine what kinds of foods and exercise routines are best recommended for long duration space flights. The ISS can help us learn how to grow food in space and even reproduce as humans.

The station has also had a questionable past which has been marred by several accidents. There was the Falcon 9 incident which failed to launch which destroyed a cargo space ship. That was also the third mission to fail in less than a year. That series of mishaps raised questions about NASA's ability to keep the station adequately supplied.

NASA's not quite sure what kinds of missions it will pursue in the future which completely depends on how much money it has in its ever shrinking budget. There is a lot of speculation about a NASA mission to Mars or a NASA built floating space station above Venus, both of which will almost certainly include the International Space Station. Why? Probably because the space station is already here, it's already in use and it could be compared to a used car. Better to fix it, save money and be familiar with what you've got, than to spend more money on something new that is unproven.

Frustrating Speed Limit!

Of all the limitations placed on space exploration the most frustrating has to be our inability to travel fast enough to traverse significant distances. The space craft we build are far too slow to consider anything but a trip to the Moon or Mars at best. While some of those crafts travel at 25,000 thousand miles an hour, an impressive speed on Earth but still a far cry from 186,000 miles per second which is the speed of light. For starters engineers need to construct a space craft that is capable of traveling at least 7 times faster than our fastest vehicles. Problem is, our current crafts are the best we can do and there's nothing in the near future to suggest any major breakthrough.

The argument that technology enabled humans to travel faster than the speed of sound does not apply to space travel. At 767 miles an hour the sound barrier is a lot different than the 186,000 mile per second speed of light barrier which is also entirely too slow for practical use in space. It is like trying to compare the notion of manifest destiny to space exploration. Exploring the plains of early North America provided water to drink, fresh air to breath and animals to eat. Outer space provides none of those essentials only instant death.

It you cut through all of the complex explanations of why we can't travel at light speed, the most common sense reason seems to be ignored which is light doesn't weigh anything. So if light has no weight then it would also seem that light has no mass and without weight and mass, would it not be easier to achieve amazing speeds? Do scientists consider these obscure ideas? Do they consider looking at things like quantum mechanics where subatomic particles move at amazing speeds? Later in the section we will take a brief look at the subject.

So what is light? According to scientist it is some sort of a wave in motion thought to transfer energy from one point to another without actually moving any material. Scientifically speaking a light wave is an electromagnetic disturbance that is made up of rapidly moving electric and magnetic effects that transport electrical charges from the sun and other bright objects to electric charges in the human eye, and when we see that light we become aware of its energy that we just received.

Many scientists say the faster an object moves, the greater the mass of that object becomes. In other words, the faster it moves the bigger it becomes. That is one part of the Theory of Special Relativity which posits, as something speeds up, its mass increases as compared with the rest of its mass and the more energy that is put into moving the object only increases the size of the object and not its speed.

As far as scientists think, nothing can travel faster that the speed of light, at least that's what they think. According to Albert Einstein light travels 186,000 miles an hour in a vacuum. Light travels slower outside of a vacuum, for example light travels slower when it passes through glass or clouds. So if light travels slower outside of a vacuum is there an environment where light actually travels faster than it does in a vacuum? There's another bizarre possibility that might send scientists searching in new directions when it comes to increasing speed for significant space travel.

A look at particle accelerators like the Large Hadron Collider could provide some suggestions for getting up to light speed but it might not provide much motivation. It might seem that the smaller the object, the faster it can be propelled. Scientists working in the LHC have been able to accelerate sub atomic particles very close to the speed of light. But they did have some help because those electrons were already traveling at more than 90 percent of the speed of light but still a tremendous amount of more energy was needed for that extra little boost. Put another way if a rocket ship was traveling at 95 percent of the speed of light, it would be doubtful if we could come up enough extra energy needed to propel the ship just 5 percent faster. Hopefully that hypothetical situation illustrates just how much energy is needed to increase the smallest increments of speed when you are traveling at speeds approaching that of light.

Will science ever find a way where humans can travel at those speeds? Scientists like Mr. Einstein say no. Reason being, the Theory of Special Relativity, which puts forth the faster an object goes the bigger it becomes which requires more energy until the point where you just can't produce any more energy to make it go faster because of its increased size. Could the space ship of the not so distant future ever approach a fraction of the speed of light? Again Mr. Einstein would say no but others in the scientific community are not quite sure. Of course you can't look into the future but the past can give some indication of what might be ahead. Remember the scientific advancements of the past made what was once thought impossible, possible, we'll there's no reason to think that won't happen again which is an appropriate but brief introduction into something called Quantum Mechanics.

Quantum mechanics is the study of physics on the smallest of levels, going beyond the atom to the protons, neutrons and the orbiting electrons that comprise the atom. Albert Einstein's theories could not explain how these particles interacted with each other at the sub atomic level. His theories were based on how gravity could explain the behaviors of planets and galaxies but was confused by the smaller world where gravity was of no use to explain the actions of atoms and smaller elements. At the sub atomic level gravity is too weak to be relevant and there's no order on this small scale, only chance and randomness, with no way to predict a certain outcome, one can only predict the chances of a certain outcome. At the subatomic level uncertainty is the only certainty. Could part of quantum mechanics be applied to the concepts of light speed and maybe later even to sight speed? It seems part of the reason why light travels so fast is light doesn't weigh anything. An atom weighs nearly nothing at 12 atomic units. It takes about 600 sextillion atoms to make one gram. That is a 6 followed by 23 zeros. Could part of the secret to light speed and eventually to sight speed be found at the sub atomic level? This much is certain; we've already have learned a lot about the universe by examining its smallest parts.

At present scientists don't know how to make the kind of propulsion systems required for the task at hand but the future space craft might not be propelled but instead pulled through space and time. Technology and nuclear fusion might advance in the next century to the point where space travel reaches a new dimension. Kind of like when the Wright Brothers started flying at 30 miles an hour to that first shuttle flight at 17,000 miles an hour in the time span of just 78 years.

Current State!

Even though we can't travel at the speed of light we can currently communicate that fast. With the use of radio waves humans can send messages to distance locations at about 186,000 miles per second or roughly 670 billion miles an hour which is approximately 6 trillion miles a year. Within radio waves probably lies our greatest hope of communicating with intelligent alien life somewhere out there in the galaxy. Astronomers are fairly sure there is no such life inside the solar system.

Scientists recently sent a message that was coded to a globular cluster. A globular cluster is usually a cluster of large, old stars that are thought to be on the outskirts of the galaxy. The message was sent to the globular cluster M13 is in the constellation Hercules. Problem is that is a long way off, estimated to be 48,000 light years out in space. So on the chance that the message was received and returned by the same method, it would take another 48,000 light years before the message was answered. And in that time there is no guarantee humanity will still be here.

As mentioned earlier space exploration is bound by many things including a budget. Because it's cheaper to receive than to send, scientists concentrate a majority of their efforts on receiving messages than sending them. Making a difficult task even more difficult is the element of timing. There's the possibility that we might have already received a message sent by radio waves from another civilization but we might have been looking in the wrong direction for that split moment that the message was receivable. Or we might have been looking in the right direction but had a momentary lapse of attention and missed the message all together. Or we might have looked in the right direction at the right time but were on the wrong frequency. The number of possible errors is too long to list here but suffice it to say the task is extremely daunting.

Truth of the matter, we don't even know if we can really receive messages, we just think we can because no such message was ever confirmed. The closest we got was with that mysterious dispatch called The Wow Signal. It was recorded in 1977 on a narrow band radio signal and despite repeated attempts over the years the signal was never detected again and its source was never confirmed.

With present technology we really can't send probes to other planets in our search for life because it would be too expensive. There is also the consideration that data sent back from such probes would be too insignificant to justify the cost and time associated with the effort. One notable probe sent far out into space was named appropriately enough Pioneer 10. It was the first to enter the Asteroid Belt, some 350 million miles from Earth. Pioneer 10 was launched in 1972 and was the first craft from Earth to fly by Neptune more than 10 years later in 1983. Neptune is some 2.7 billion miles away and is near the outer reaches of the solar system. The probe cost an estimated 350 million dollars and by the turn of the century it could be more than 7 billion miles from Earth. By January of 2003 the probe had been on the job for about 30 years and sent what was likely its last message to Earth. It was a frail, weak signal that scientists were barely able to detect. Three years later scientists made one last attempt to receive another signal from Pioneer 10 but to no avail. So despite all of its success the probe, to the best of our knowledge, never sent back any data or information that would indicate it found intelligent alien life or even if intelligent alien life exists somewhere out there.

If there are maybe 500 billion galaxies in the universe that means, there could be 50 sextillion planets out there that might be habitable.

A more manageable figure might be 50 billion planets in our solar system and about 500 million of them could be in the habitable zone. So the chances of finding life are pure speculation at best but could be something like one in several million.

One popular calculation method takes into account the number of stars in the galaxy and then the number of those stars that have planets orbiting them. The method then goes on to consider the number of those planets that are in the habitable zone, then factor in the number of planets that actually have life on them. From there it goes on to factor in how long it takes intelligent life to develop. After everything is considered there might be only one planet out there, Earth or there might be more than a million. But for people who are guided only guided by fact there is only Earth and nothing else out there.

Mother Earth!

We would be remiss not to mention so many of the amazing phenomena in the universe without writing more about Earth, the most unique planet that we know of. Again, Earth is so rare it might have been created by accident. Our planet seems to have just the right combination of all the rare essentials required for our survival. It is just the right distance from the sun, it has just the right combination of water and land, the right amount of molten material beneath the surface, the right mixture of chemicals to make up the complex atmosphere, enough luck to avoid a cosmic disaster and enough time for humans to evolve.

Unlike most planets Earth is living in every sense of the word. Its surface can be compared to human skin, constantly growing and renewing. Part of the process is due to tectonic plates, which are giant underground slabs of rock, 50 to 250 miles thick and can stretch for thousands of miles in every direction. All of Earth's water and land rest on these plates. The plates constantly move, shift and renew the surface. The planet is about 4.5 billion years old and as soon as it cooled, rocks formed and have been recorded in the geologic record which are embedded in these tectonics plates. At one point there was just one giant continent instead of the seven we now have. That one massive continent was called Pangaea, thought to exist some 250 million years ago. Movement of the plates is also called continental drift and is thought to create volcanoes, earth quakes and mountains ridges. When the plates suddenly break apart they can cause earthquakes, when the plates collide with each other they push the surface up and can create mountain ranges and when they move in certain directions they can disrupt giant underground reservoirs of magma or liquid rock which can cause volcanoes.

Continental drift can also explain why identical rocks that are 200 million years old are found several thousand miles apart. An example would be certain rocks in Africa are identical to rocks in North America. Those land masses used to be joined together but after the continental drift they are now thousands of miles apart.

The current atmosphere is just another marvel of evolution because the early atmosphere would have killed us. It was noxious, it was toxic, it was searing hot. Advanced life could not have survived with those high concentrations of hydrogen, nitrogen and ammonia and the searing heat, combined to provide the death knell for even simple life like bacteria. It was not until about 3.8 billion years ago that Earth's atmosphere began to cool to the point where land formed and later rain fell heavy enough to form oceans. The oxygen expansion did not occur until about 2.4 billion to 500 million years ago. Before that before the life giving oxygen was basically non-existent. 500 to 250 million years ago the atmosphere started to become stable enough to allow complex life to start to develop. During the evolutionary change large amounts of air became present in the atmosphere. Due to earth's gravity about half of all our air is packed in the first four miles above the ground. About six miles up and humans would suffocate. The ozone layer can be found about another mile up which protects us from the sun's harmful ultraviolet rays. As mentioned earlier about 62 miles up is where space begins.

Sometime between 250 and 60 million years ago animals began to evolve. The ice age was 35 million years ago and the global warming problem began from 1880 to the present. There really is nothing like the Cosmic Calendar to put things into perspective. It was shown in the TV series Cosmos: A Space time Odyssey, and compresses the age of the universe, 13. 8 billion years old into a one calendar year. The beginning or the Big Bang occupies January 1 at midnight and the current moment occupies the last second on the calendar at December 31 at midnight. On this scale a second contains about 434 years, one hour contains about 1.5 million years and a day contains almost 38 million years and one month is about one billion years. Primitive life first evolved on Earth September 21st, simple animals appeared December 7th, dinosaurs showed up December 25th, primates on December 30th, hominids or primitive humans on the next day and in the last minute came the first recorded civilization with the Sumerians.

During the last seconds of the year came the European Renaissance and within the very last second came us, the American Revolution, the abolition of slavery, two world wars and the moon walk. Within that last second on the calendar represents about the last four hundred years and the first second was the beginning of the universe. Humans have apparently beaten very long odds and had very little time in which to do so. Does the Cosmic Calendar emphasize how fortunate we are to live in the here and now?

The only endeavor in which humans have collectively worked together since the beginning of history might very well be Astronomy. Think how far we have come since the infancy of astronomy, a distance much too far to ever have been traversed by one or a handful of individuals. The gains in astronomy are only made possible through the collective efforts of successive generations, with current knowledge built upon the knowledge that preceded it and astronomy might be the only science where amateurs like you and I can join the conversation and make significant contributions that can be immensely rewarding.

So much of astronomy is theory but is that a reason to believe scientists are leading us in the wrong direction? History has shown that when given enough time science will provide answers to the most perplexing questions. We might already have the answer to that age old question of our origin and we just don't like it. The human character is rarely satisfied with anything. Science states we came from the Big Bang and from star dust, the universe is in us and we are in the universe, even the very iron in our blood came from stars. So that's our origin and until we learn more that's all there is. The brief but encompassing narrative might also suggest how small humans are in the context of the universe and our short life spans make us even more insignificant. This understanding can provide humility, satisfaction and a better appreciation for all the things making up the here and now.

THE END